Electronic Projects in the Home

Other Constructors Projects Books

Electronic Projects in Audio
Electronic Projects in the Car
Electronic Projects in Hobbies
Projects in Radio and Electronics
Electronic Game Projects
Electronic Projects in Music
Electronic Projects in the Workshop

Electronic Projects in the Home

Owen Bishop

Series Editor Philip Chapman

Newnes Technical Books

The Butterworth Group

United Kingdom — Butterworth & Co (Publishers) Ltd
London: 88 Kingsway, WC2B 6AB

Australia — Butterworths Pty Ltd
Sydney: 586 Pacific Highway, Chatswood, NSW 2067
Also at Melbourne, Brisbane, Adelaide and Perth

Canada — Butterworth & Co (Canada) Ltd
Toronto: 2265 Midland Avenue, Scarborough, Ontario M1P 4S1

New Zealand — Butterworths of New Zealand Ltd
Wellington: T & W Young Building
77–85 Customhouse Quay, 1, CPO Box 472

South Africa — Butterworth & Co (South Africa) (Pty) Ltd
Durban: 152–154 Gale Street

USA — Butterworth (Publishers) Inc
Boston: 19 Cummings Park, Woburn, Mass. 01801

First published 1979

British Library Cataloguing in Publication Data

Bishop, Owen
Electronic projects in the home.
1. Electronic apparatus and appliances – Amateurs' manuals 2. Household appliances, Electric – Amateurs' manuals
I. Title
621.381 TK9965 78-40663

ISBN 0 408 00346 4

Typeset by Butterworths Litho Preparation Department

Printed in England by William Clowes & Sons Ltd., Beccles and London

Contents

1

Moisture Detector

The water tap to which I connect my lawn sprinkler is out of sight of the lawn, and it is almost impossible to adjust the rate of flow by listening to the sound made by the water rushing from the tap into the hose. Instead there is a lot of dashing about between tap and lawn, and much exasperation, until the rate of flow has been adjusted to cover an adequate area of lawn, yet not drench people walking past outside the garden fence. Now, with the arrival of the moisture detector, all this chaos is at an end. The probe is placed at the edge of the area that is to be watered, and the tap is opened gradually to increase the rate of flow. The alarm sounds as soon as the first few drops of water fall on the probe plate. Then I know that the sprinkler is correctly adjusted, even without going outdoors to look.

The moisture detector is triggered by a single drop of water, so it can also be used to warn you to bring in the washing when it starts to rain, or to tell you when a bath or water tank is full or overflowing. It also works with a cloud of fine droplets, so can be used to tell you when the kettle is boiling, by placing the probe where the steam will play on it. If the probe plate is cool, the condensation from your breath will make it operate. Even a slightly damp finger pressed against the probe plate will result in a sound from the detector. The pitch of the sound varies according to the amount of water present, so you can gauge how hard it is raining, or how vigorously the kettle is boiling. By varying the pressure of your finger on the plate you can even play a tune!

The circuit

The detector is based on a single unijunction transistor (Fig. 1.1). This can be thought of as a bar of semiconductor material with base contacts

(b_1, b_2) at either end. The resistance of the bar is approximately 5 to 10kΩ, so little current flows from b_2 to b_1. If moisture is present on the probe, there is an electrical connection through the moisture between the two probe electrodes; current flows through the probe and through R1, gradually charging C1. As the charge on C1 increases, the potential of the emitter (e) of the transistor gradually rises. When the

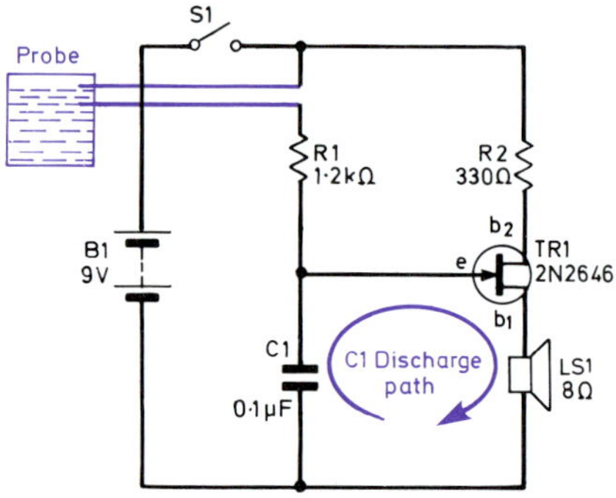

Figure 1.1

Circuit diagram of the moisture detector

potential at e reaches a certain value (about 6V), current begins to flow into the bar from e, and this has the effect of reducing the resistance of the bar between e and b_1. Consequently a large current flows from e to b_1, and C1 is rapidly discharged. Current flows until the potential at e has fallen to about 2V. Thus, as long as there is water on the probe, C1 is alternately charged to about 6V and discharged to about 2V. An intermittent current flows from e to b_1 and out of the transistor, through the loudspeaker to the negative terminal of the battery. In other words, this circuit is a simple type of *oscillator.* the rate at which C1 charges and discharges depends on the amount of current entering it through the probe and R1. This varies according to the amount of water present on the probe. With suitable amounts, and a suitable value for R1, the capacitor charges and discharges several hundred times per second. The associated intermittent current from b_1 produces a musical note (with a frequency of several hundred hertz) as it passes through the loudspeaker LS1. As the amount of water increases, the probe resistance is reduced, causing C1 to charge and discharge more frequently, and so producing a note of higher frequency (or pitch). With a very small amount of water on the probe, the charging current is minute; under these conditions it is possible to hear the individual clicks each time C1 is discharged through e, b_1 and the loudspeaker.

Construction

Building the detector presents no problems. Fig. 1.2 shows how the components may be arranged on a piece of 0.15in matrix stripboard. The off-board wiring is shown in Fig. 1.3. The probe may be plugged into a coaxial socket, as shown, or can be wired permanently to the circuit. Any twin flex can be used to connect the probe to the circuit.

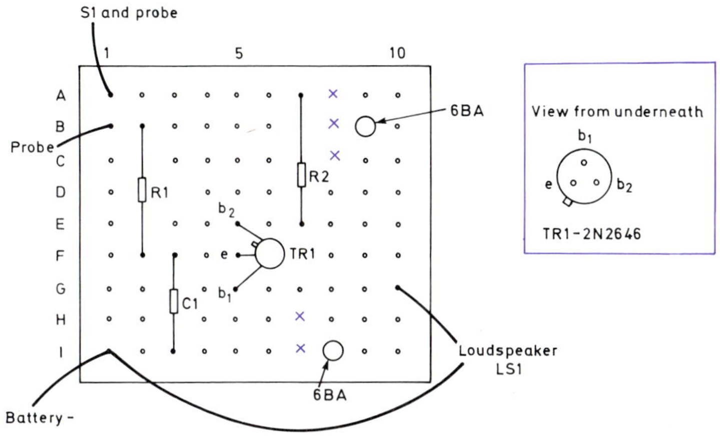

Figure 1.2

Layout of components of the moisture detector

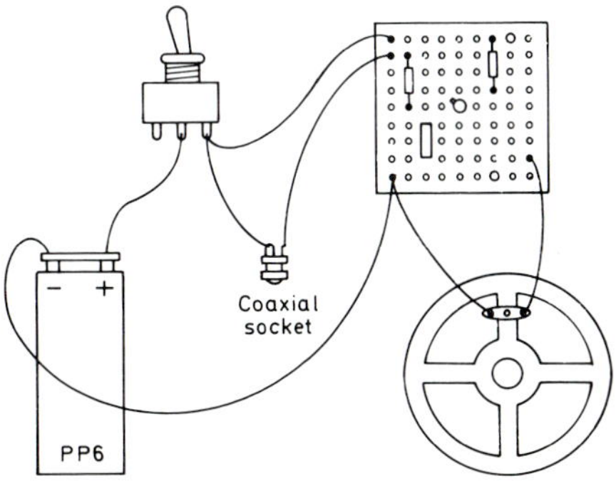

Figure 1.3

Interconnections between the circuit board and other components of the moisture detector

The probe itself is constructed from stripboard with alternate strips connected by wires (Fig. 1.4). Any spare piece of board may be used.

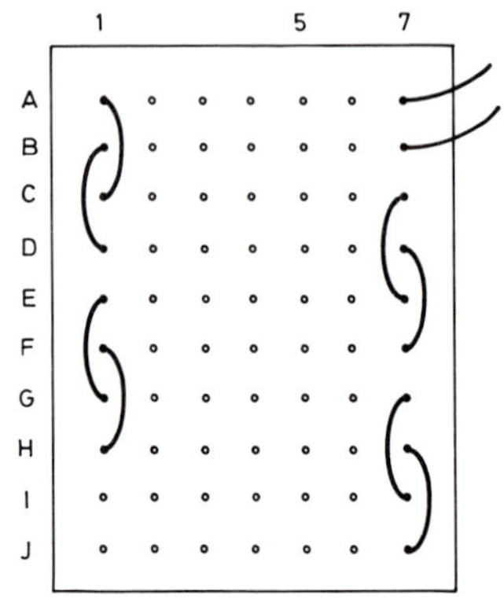

Figure 1.4

Wiring the scrap of stripboard to make a moisture probe

Figure 1.5

The completed moisture detector. The lead to the probe plate can be much longer than that shown here, if required

A plastics or metal box may be used to house the circuit (Fig. 1.5). The area in front of the speaker is drilled with a number of small holes in regular pattern. Alternatively, a large circular hole may be cut and covered with a grille of wire mesh or stiff fabric. The speaker is fixed in position by three bolts which pass through holes drilled in the front panel. A solder tag is fixed to each bolt, and the end of the tag is turned inward to grip the rear of the loudspeaker frame. The circuit board may be held in place by bolting it to one of the speaker bolts, and to another bolt fixed to the front panel.

Using the moisture detector

When the device is assembled, switch on, sprinkle a few drops of water on the probe, and listen to the note of varying pitch that comes from the speaker. If no sounds come forth, check all soldered joints to ensure that you do not have any 'dry' joints. Also check that there are no unwanted strands of solder connecting adjacent strips on the circuit board. A hair-thin strand can easily present a circuit from functioning properly.

When all is working properly, you are ready to use the detector in a wide variety of applications. Perhaps, with a suitably moistened finger, you will be the first person to compose a Concerto for Moisture Detector!

Table 1.1. Components list for moisture detector

Resistors	
R1	1.2kΩ
R2	330Ω
Both carbon film, $^1/_4$ or $^1/_8$W	
Capacitor	
C1	0.1μF polyester or polycarbonate
Semiconductor	
TR1	2N2646 unijunction transistor
Miscellaneous	
0.15in matrix stripboard, 9 strips × 10 holes (see Appendix, p. 87); 0.15in or 0.1in stripboard scrap for probe plate; case; SPST switch, any type; miniature 8Ω loudspeaker; battery connector; miniature coaxial socket and plug (optional); 6 B.A. bolts (4 off); 6 B.A. nuts (8 off); 6 B.A. shake-proof washers (8 off); 6 B.A. solder tags (3 off); terminal pins (ıor 0.15in board) (4 off); connecting wire; solder	

2

Temperature-operated Alarm

The triggering of this device depends on the action of a thermistor. This is a bar, disc or bead of semiconducting material with the special property that its electrical resistance decreases appreciably as its temperature increases. The variation of its resistance is used to trigger a circuit to switch on an alarm device whenever the temperature exceeds a given level. We can arrange for the alarm to be set off when the greenhouse becomes overheated, when a saucepanful of milk is about to boil over, or when the house catches fire! The circuit can be modified to operate in a reverse manner, so that the alarm is triggered when the temperature falls *below* a given level, making it useful as a frost detector. This and several other modifications of the circuit are described later.

The circuit

The thermistor (R1) and the potentiometer (VR1) are connected to act as a potential divider (Fig. 2.1). The potential at point A is between 0 and 9V, depending on the setting of VR1 and the temperature of R1. Note that though these two components *appear* close together in the diagram, they may in practice be far apart. The thermistor may be remotely situated and joined to the remainder of the circuit by a lead several metres long. For example, it might be in the greenhouse, with the main part of the circuit placed indoors.

As the temperature of R1 increases, its resistance decreases. The effect of this is to reduce the potential at A. (Assuming that the setting of VR1 is not altered). In the un-triggered condition, the potential at A is sufficient to cause a current to flow through R2 to the base of TR1, causing collector current to flow through R3, TR1 and R6. In this condition, the greatest voltage drop is across R3, so point B has potential of only a fraction of a volt above zero, and the potential at point C is likewise only slightly greater than zero.

When the potential at A falls below about 1.25V, the base current to TR1 falls below the value required to maintain the collector current.

Figure 2.1

Circuit diagram of the temperature-operated alarm

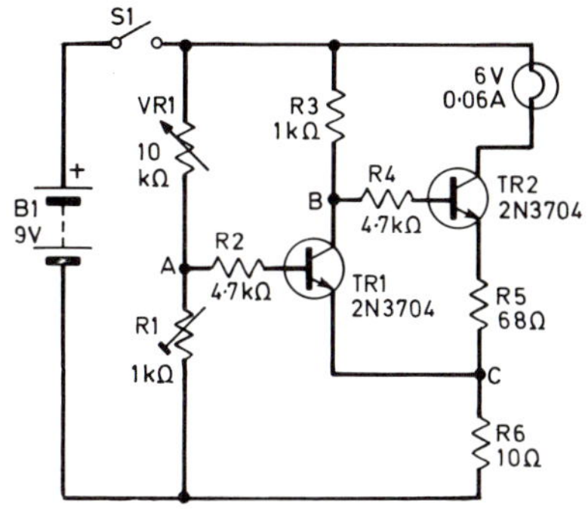

TR1 becomes switched off. Thus for a relatively small fall in potential at A, the potential at point B rises rapidly. As the potential at B rises, a base current begins to flow through R4 to TR2, turning it fully on. A large collector current flows through the lamp, and through TR2, R5 and R6. The lamp lights; the alarm has been triggered.

Once the circuit has been triggered, the lamp remains lit, even when the temperature of R1 falls slightly. This prevents the lamp being flashed on and off in haphazard fashion if the temperature is fluctuating around the set level. The explanation of this action is as follows. An adequate base current flows through TR1 only when there is a sufficiently great potential difference between A and C. When the circuit is in the untriggered condition, the potential at C is very close to zero, and the potential difference between A and C is about 1.25V. As A falls below this, the base current becomes too small to keep TR1 switched on. Once the circuit has been triggered, a large current is flowing through the lamp TR2, R5 and R6. This means that the potential at C is raised by a fraction of a volt above zero. To obtain a base current sufficient to switch TR1 back on again the potential at A must be raised 1.25V above the new, higher potential at C. In practice, it must be raised to about 1.275V. To achieve this and so switch off the alarm, the temperature of the thermistor must be less than it was when the alarm was switched on. For example, if VR1 is adjusted so that the alarm is triggered at 25°C, the temperature of the thermistor must fall to about 24.3°C before the potential at A rises to 1.275V and the alarm is switched off. By using a circuit of this type we have combined sensitivity to temperature level with a clear-out triggering action.

Construction

Layout of the components is shown in Fig. 2.2. Construction presents no problems. The circuit board and battery may be housed in any convenient plastics or metal case. If it is required to be able to adjust the temperature setting frequently, VR1 should be a panel-mounting

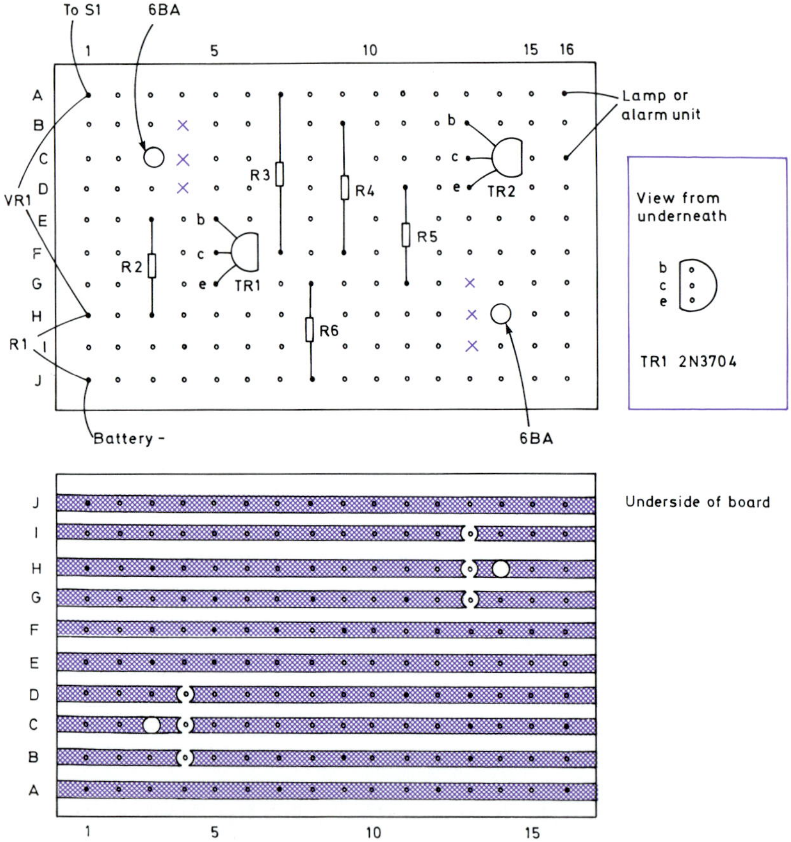

Figure 2.2

Layout of components of the temperature-operated alarm

potentiometer (volume control). If the setting is to be done once and for all, VR1 can be a miniature preset potentiometer, mounted on the circuit board. When all components are connected, switch on S1. Alter the setting of VR1. As you increase the resistance of VR1, the potential at A will fall and, at a certain setting of VR1, the lamp should light. It will then be necessary to turn VR1 a little way back to extinguish the lamp again. If VR1 is adjusted so that the lamp *just* does not light, and you then grip R1 between your fingertips, the warmth from your fingers should be sufficient to trigger the circuit. If the circuit does not respond in the way described, check wiring, and solder joints, and look for short circuiting strands of solder.

To adapt the circuit as a frost detector, or for any other application that involves triggering as temperature falls, simply interchange VR1 and R1. You may also wish to have an audible alarm instead of or as well as a lamp. For this purpose you can use the unijunction oscillator described in Project 1. Build the circuit on the board as shown in Fig. 1.2, omitting the connections to probe and the probe terminal pin (B1), and solder the upper end of R1 in hole A2 instead of hole B2. The output pin (A16) of the trigger circuit is then connected to pin A1 of the oscillator, and the other output pin (C16) is connected to pin I1 of the oscillator. Both circuits can, of course, be constructed on one piece of board, in which case strip A is left unbroken as a connection between the two circuits. If you wish to employ both lamp and sound alarms, the lamp and oscillator may be wired in parallel with each other.

Table 2.1. Components list for temperature operated alarm

Resistors	
R1	Thermistor, Siemens type K 164, or similar, 1kΩ at 25°C
R2	4.7kΩ
R3	1kΩ
R4	4.7kΩ
R5	68Ω
R6	10Ω
Potentiometer	
VR1	10kΩ linear, volume control type, or preset
Semiconductors	
TR1, TR2	2N3704
Miscellaneous	
Stripboard, 0.15in matrix, 10 strips by 16 holes; SPST switch, any type; knob for potentiometer, lamp 6V, 0.06A, and socket; battery connector, 6 B.A. bolts, nuts and washers; terminal pins for 0.15in board (5 off); case; connecting wire; solder	

As an alternative to the oscillator of Project 1, you may use the slightly more complicated oscillator from Project 8. This has a more melodious sound; it is pleasanter but less alarming than the simple oscillator circuit.

The trigger circuit can also be used to actuate a relay. This in turn can switch on (or off) mains-powered devices such as a mains lamp, a heater, a fan (for greenhouse cooling), or any other device driven by an electric motor. See Project 5 for details of how to connect and install a relay. In adapting the circuit for other purposes you may connect two, three, or even more lamps, alarms and relays across pins A16 and C16, but remember that the *total* current passed by these devices must not exceed 800mA, the maximum rating of TR2. If heavy currents are to pass through TR2 for prolonged periods, some form of heat sink may be required. You may also find it worthwhile to power the circuit from a 9V power-pack (Project 12, for example) instead of a battery.

3

Flashing Messenger

If you are leaving a message, and want to make sure that it is noticed, use the Flashing Messenger to attract the attention of the person for whom the message is intended. The Messenger consists of a small flat box (such as a tobacco tin) with a decorative picture or design on one side and a smooth writing surface of a plastics material (e.g. Formica) on the other. The picture incorporates a small lamp which flashes to attract attention. Then the recipient turns the box over to read the message written on the plastics surface. Alternatively, the Messenger can be used as an electronic 'knot-in-the-handkerchief', to remind yourself of something you must not forget to do. Leave it flashing beside the telephone, to remind yourself to make that important call; leave it flashing on the cooker, to remind yourself to take out the cake when it is baked; leave it flashing on top of the TV set, to remind yourself to switch on that programme you must not miss. The Messenger has a multitude of uses in the home, and many in the office too.

The circuit

The circuit is extremely simple to build, being designed around a single integrated circuit. In this book we are not concerned with the dozens of electronic devices incorporated within the integrated circuit (or i.c. for short), neither are we concerned with exactly how it works internally. We are adopting the 'black box' approach, and are concerned only with what the i.c. can *do*.

The i.c. used for the Messenger contains two identical logic circuits, which we use as *inverters*. Their action is simple. If the input to an inverter is 'high' (meaning that the voltage at the input terminals is at or close to 4.5V), its output is 'low' (meaning a voltage very close to zero). Conversely, if its input is 'low', its output is 'high'. In logic circuits, inputs and outputs are considered as either high or low – no intermediate state is recognised. The two inverters are connected to make an oscillating circuit, or *astable multivibrator.* In Fig. 3.1 the inverters are represented by triangles, with a small circle at one point (the output) to indicate that they have an inverting function. The diagram shows that the circuit can be in one of two states. In state A, inverter 1 has its output high (indicated by colour because its input is low). In this state the potential of *both* plates of capacitor C is high at first, and thus the input to inverter 2 is high also. This causes the output of inverter 2 to be low. Note that resistor R has a low potential at one end, and a high potential at the other. This causes a current to flow through R, in the direction indicated by the arrow.

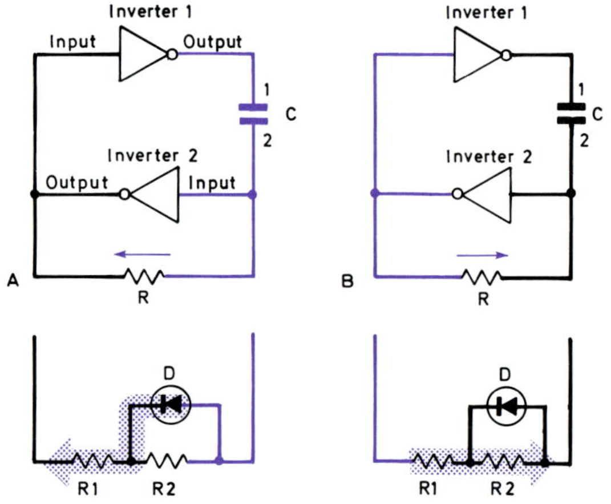

Figure 3.1

The operation of an astable multivibrator, built from two inverters, a resistor and a capacitor

As the current flows, the potential of plate 2 of the capacitor gradually falls. The higher the value of R and the greater the capacitance of C, the more slowly the potential falls. Eventually it will have fallen to a level low enough to constitute a low input to inverter 2. At this instant the output of inverter 2 suddenly changes to high. This makes the input to inverter 1 high, so its output goes low. This brings down the potential of both capacitor plates to low. The circuit is now in state B.

In state B we again have high potential at one end of the resistor and low potential at the other. Again a current flows, but in the opposite direction. Gradually plate 2 of capacitor C is charged, until the input to inverter 2 becomes high and the circuit reverts to state A again. In this way the circuit oscillates between state A and state B at a frequency depending on the values of R and C. Any point in the circuit alternates between high and low potential at this frequency.

In the circuit shown in Fig. 3.1, state A and state B last for equal periods of time. The lower diagrams show how we may replace R by

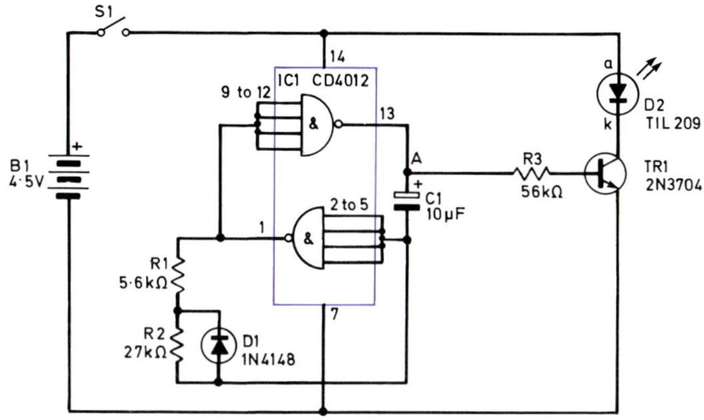

Figure 3.2

Circuit diagram of the flashing messenger

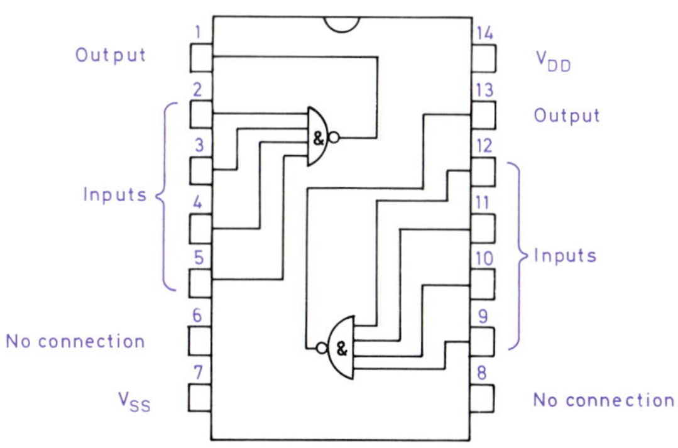

Figure 3.3

Pin connections of the CD4012 i.c.

two resistors, R1 and R2, and a diode D. When the circuit is in state A, R2 is bypassed by D. If R1 has a low value the potential at the input of invertor 1 falls quickly, so the length of time spent in state A is short. When the circuit is in state B, the current must pass through R1 *and R2,* in series. If R2 has a high value, the potential at the input to inverter 1 rises very slowly, so the length of time spent in state B is long.

The multivibrator is constructed from the two logic gates contained in the CD4012 integrated circuit. These gates are of the type known as 4-input NAND gates. They give a 'low' output only when *all four* inputs are 'high'. If any one or more of the inputs are 'low' the output is 'low' too. To obtain the inverter function with this type of gate we connect all four inputs together, as shown in Fig. 3.2. Fig. 3.3 shows the pin connections of the i.c. The multivibrator circuit comprises the two NAND gates of the i.c., together with C1, R1, R2 and D1. Compare Figs. 3.1 and 3.2 to work out which parts correspond. The current required to operate the gates is supplied to the i.c. through pins 7 and 14. The CD4012 belongs to the complementary metal oxide silicon family (CMOS), in which the switching actions are performed by MOS field-effect transistors. These require very low currents, which is advantageous for a device which is to be battery-powered and may need to be left switched on for several hours at a time. If we think of an electric current as it really is – a flow of electrons from the negative battery terminal to the positive battery terminal – the electrons enter the i.c. via pin 7. This is the *source* of electrons for all the transistors in the i.c.; the potential applied to this terminal is V_{SS}, and is normally 0V. The electrons leave the i.c. via pin 14. This is the pin through which the electrons *drain* from the transistors; the potential applied to this terminal is V_{DD}, and can be between + 3V and + 15V. In the Messenger circuit V_{DD} is + 4.5V, conveniently supplied by the flat 3-cell battery which fits comfortably into a flat tobacco tin.

As the circuit oscillates the potential at point A (Fig. 3.2) alternates between 0 and 4.5V. When it is high (4.5V) a base current flows through R3 to TR1. The transistor is switched on; a collector current flows through it via the light-emitting diode D2. This emits light for as long as the multivibrator is in state A. When the potential at point A is low, there is no base current and consequently there is no collector current, so the LED is dark for as long as the multivibrator is in state B. During state A the current required by the circuit is about 20mA, almost all of which is flowing through the LED to produce the light it emits. During state B the current required is only 1mA, the small amount that is needed to power the i.c. The values of R1 and R2 have been chosen so that the multivibrator spends about six times as long in state B as it does in state A. The LED flashes briefly, with relatively long dark periods between flashes. In this way we restrict current consumption and prolong battery life without reducing the eye-attracting effectiveness of the flashing LED.

The prototype was built to fit inside a rectangular tobacco tin (Fig. 3.4) but any plastics or metal container of similar size and shape would be suitable. The long narrow circuit board (Fig. 3.5) carried the main components at one end, the remainder of the board bearing the two drawing pins that act as battery contacts.

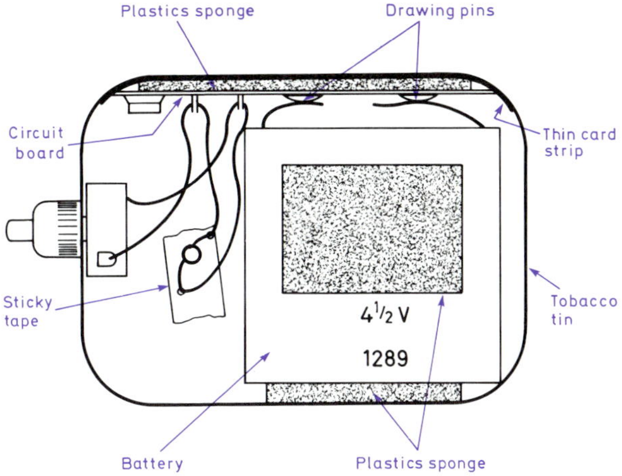

Figure 3.4

Interconnections between the circuit board and other components of the flashing messenger

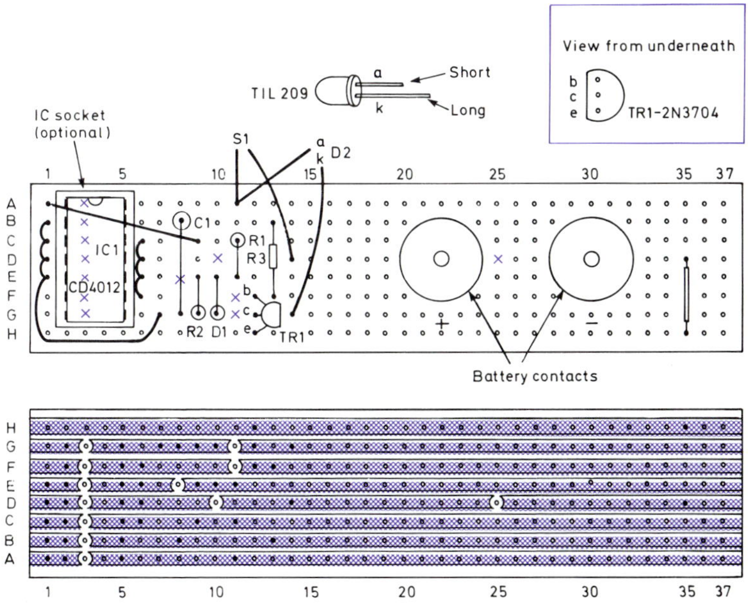

Figure 3.5

Layout of the circuit board of the flashing messenger. The leads of the TIL209 light-emitting diode are identified by their relative lengths

Construction

When handling CMOS, certain precautions must be observed (see Appendix) and beginners may find it easier to build the circuit with an i.c. socket, and insert the i.c. only after construction is complete and the positioning and soldering of all other components has been carefully checked. Yet the cost of an i.c. socket is more than the cost of the i.c. used in this circuit, and with careful observance of the precautions there should be no problem in dispensing with a socket and soldering the i.c. direct to the board. It is best to build the entire circuit before testing it. Take care when pushing the drawing-pins through the holes in the board. Thick pins may force the copper strip away from the board, causing breaks in the strip which may be very hard to see. Solder the pins to the strips, then cut off their points. Mark the board with bold plus and minus signs, so that no mistake can be made when inserting the battery.

Test the board wiring before inserting the board in the case. For testing, connect the battery temporarily, using leads with crocodile clips. As soon as S1 is turned on, the LED should flash regularly and brightly, about 1 flash per second. If it does not flash, check that the LED has been connected with correct polarity. Recall that this is a diode, through which current can flow only from anode a to cathode k, and not (except for a minute leakage current) in the reverse direction. Incidentally, *never* attempt to test an LED by simply touching its leads to the terminals of a battery. The large current passed will burn it out instantly – you may not even see the flash! When testing an LED with a battery of 4.5 or 6V, always connect a resistor of 150 to 270Ω in series with it.

When inserting the board in the case, place a strip of card along the inside of the case so that the ends of component leads on the reverse side of the board can not be short circuited through the case. This is not necessary if a plastics case is used. Pieces of plastics sponge of suitable size are used to wedge the battery in position and to ensure that its terminal strips make good contact with the drawing pins. A piece of foam placed on top of the battery helps hold the battery in position when the lid is in place. Only one hole needs to be drilled in the case – to take the pushbutton switch.

The outside of the case is decorated on one side with a suitable picture (Fig. 3.6). Use a post-card, greetings-card or (as in the prototype) cut a picture from a catalogue. The LED can be placed in any position to suit the picture or design – in the prototype the LED is a flashing lapel-badge. The picture and the rectangular piece of white Formica are fixed in position by contact adhesive.

Messages may be written on Formica in pencil or felt-tip pen (Fig. 3.7). They may be erased by using an ordinary pencil eraser or by wiping with a damp cloth.

Figure 3.6

The flashing messenger waiting to deliver its message; the LED is on the messenger's collar, just above the largest star

Figure 3.7

The message is received by turning the messenger over

Table 3.1. Components list for flashing messenger

Resistors	
R1	5.6kΩ
R2	27kΩ
R3	56kΩ
all carbon, $^1/_4$W	
Capacitor	
C1	10μF (tantalum or electrolytic)
Semiconductors	
TR1	2N3704, or similar
D1	1N4148, or similar silicon diode
D2	TIL 209, or similar LED, any colour
Integrated circuit	
IC1	CD4012
Miscellaneous	
0.1in matrix stripboard, 8 strips × 37 holes; i.c. socket, 14-pin DIL (optional); terminal pins, for 0.1in board (three off); push-on/push-off switch, drawing pins (two off); case; scraps plastic foam; white Formica; picture; 4.5V battery, type 1289; connecting wire; solder	

4

Yoghurt Maker

Home-made yoghurt is considerably cheaper than that bought in shops, and many people consider that it has a much nicer flavour too. The Yoghurt Maker is a container which is held at constant temperature for several hours to allow the yoghurt micro-organisms to get to work on the milk and convert it to yoghurt. The recipe is given at the end of the chapter, and for the moment we shall be concerned only with the electronic aspects of the project.

The circuit

The regulation of temperature inside the container depends on the action of a thermistor (see Project 2). There is no special heating element

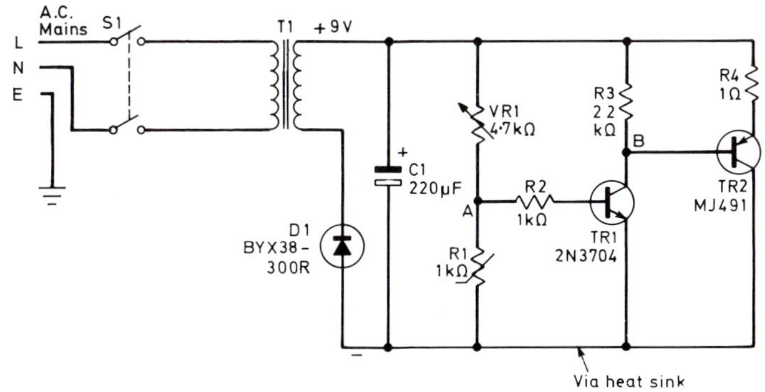

Figure 4.1

Circuit diagram of the yoghurt maker

for we simply make use of the heat given off by the electronic components themselves as current passes through them. The circuit is mains powered (Fig. 4.1), the voltage being stepped down to approximately 9V, rectified by a single diode D1, and smoothed by capacitor C1. With a rise in temperature, the resistance of the thermistor (R1) decreases, causing fall of potential at point A. This reduces the base current to TR1 which is gradually turned off. As it becomes turned off the potential at point B rises almost to 9V. Since TR2 is a *pnp* transistor, this has the effect of turning off TR2 as well. Conversely, as temperature falls, both TR1 and TR2 are turned on. The current passing through TR1 is always small – never more than about 0.4mA – since the value of R3 is high. But since R4 has a resistance of only 1Ω the current passed by TR2 may be as great as 2A. The passage of such a large current through TR2, R1 and D1 generates enough heat to maintain the interior of the container at 35°C, the temperature required for making yoghurt.

Figure 4.2

The yoghurt maker ready for use. The smaller tin at bottom left in the photograph houses the transformer; the main switch is mounted on its lid. The main circuit is on the left, but resistor R4 is on the right. There is enough room for three plastics containers in which the yoghurt is made and a smaller one in which to make the starter culture for the next batch

Any metal container of suitable size can be used to house this project. For the prototype, a biscuit tin was used (Fig. 4.2). A smaller tin was used to house the transformer. Switch S1 is mounted on the lid of this inner tin. Bolts run through the mounting tags of the transformer, the bottom of the inner tin and the bottom of the larger tin to fix all three firmly together. It is important that grommets are used wherever wires pass through the metal walls of the tins.

Most of the components are assembled on a piece of stripboard (Fig. 4.3) which is bolted to the large heatsink on which D1 and TR2 are

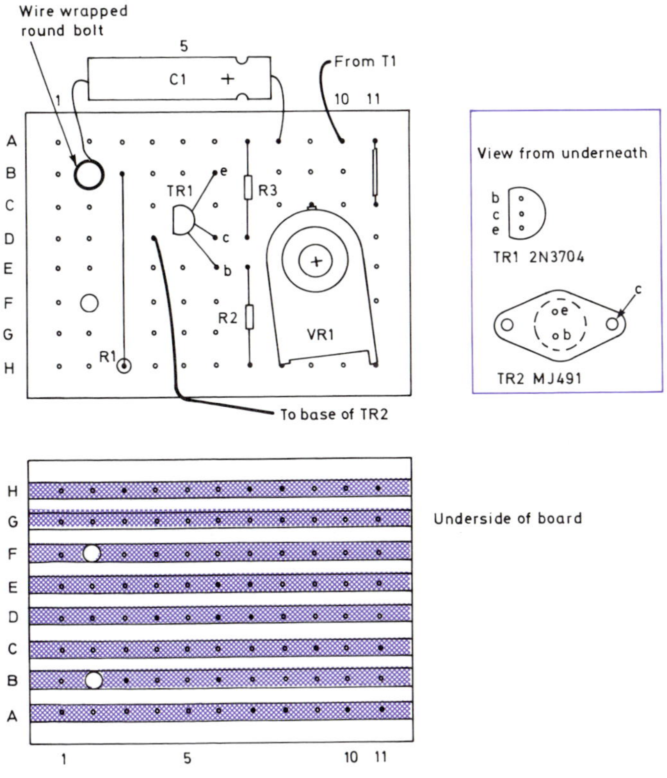

Figure 4.3

Layout of the circuit board of the yoghurt maker

mounted (Fig. 4.4). TR2 and D2 are bolted directly to the heatsink *without* the use of mica washers. The collector of TR2 and the anode of D1 are internally connected to their metal cases, so the heatsink provides the return path from the collector of TR2 to the anode of D1. This is why it is essential for D1 to be a BYX38-300R, and not BYX38-300. The latter type has its *cathode* connected to the case, so is unsuitable for this circuit. The connection between TR2 and D1 and the remainder of the circuit is made through the various flying leads and through the metal bolt which connects strip B on the circuit board to the heatsink.

To obtain uniform heating of the enclosure, it is best for R4 to be situated on the side of the container opposite from the heatsink. It is

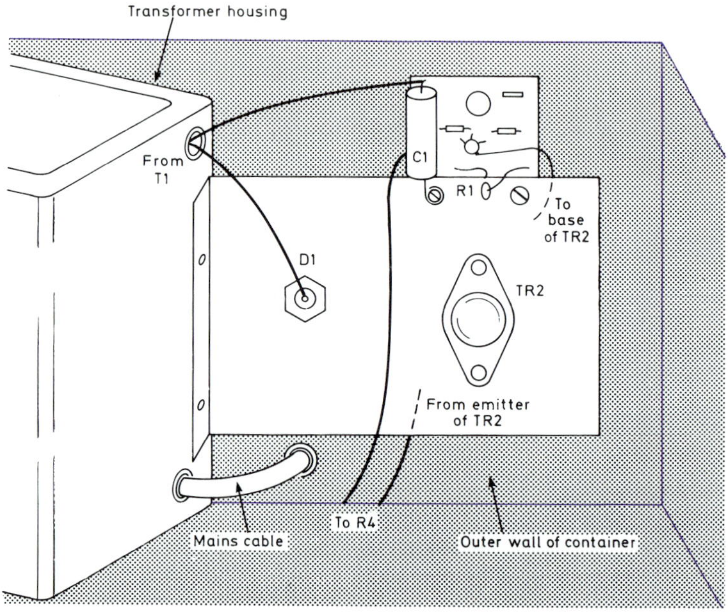

Figure 4.4

Interior view of the yoghurt maker to show the way D1, TR1 and the circuit board are mounted on the heat sink

essential for R4 to be a wire-wound resistor rated to withstand at least 2A. It can be mounted on a small piece of 0.15in matrix circuit board, which is attached to the wall of the container by a bolt.

To obtain sufficient heat insulation the container and its lid are lined with expanded polystyrene sheet. This can be cut from a sheet about 1cm thick and it is easy to wedge rectangular pieces in position on the sides and bottom of the box so that they need no special fixing bolts. The piece lining the lid is fixed by two or three bolts, with large washers to obtain a firm grip on the polystyrene without crushing it or splitting it.

When wiring is complete use a test meter to check that there is good electrical connection between the earth pin of the mains plug and the metal container and inner container. Check also that there are no short circuits between the live and neutral pins and the remainder of the circuit (with S1 switched on). Then plug the device into the mains socket and switch on. Adjust VR1 so that the potential at A is approximately 1V. TR2 and R4 will rapidly become hot, especially R4 which soon becomes too hot to touch. This is not surprising, for it is passing over 1.5A when the current is first switched on.

After a few minutes the heat from TR2 begins to affect R1, placed directly above TR2, and the potential at A falls slightly below 1V. If it does not, switch off and check wiring. If all appears to be operating correctly, place a thermometer in the enclosure and put the lid on. Inspect the thermometer from time to time, and adjust VR1 to bring the internal temperature to 35°C. To increase temperature, turn VR1 anticlockwise; to decrease temperature, turn VR1 clockwise. When the temperature remains steady for periods of half-an-hour or more, you know that the device is correctly adjusted and ready for use. You may decide that it will be improved by fitting a small guard made from wire-mesh around R4, to prevent it from touching and possibly melting the plastics containers in which the yoghurt is prepared.

How to make yoghurt

To make yoghurt you need milk and a culture of the yoghurt organisms. The culture is obtained by buying a carton of natural (plain) yoghurt. Having bought this, you need buy no more for months or even years. You need some plastics food containers, like those shown in the photograph, which have air-tight lids. Such containers are readily available. As well as the three containers for use at the table, you need a small one to hold a portion of the batch that is to be kept to start the next batch with. Bring ¾ pint milk to the boil in a saucepan. Allow it to cool until it is *just* warm to the hand (i.e. about 35°C). Add one tablespoonful of natural yoghurt to the milk and stir well. Put one tablespoonful of the mixture in the small container and divide the remainder between the three large containers. Put the lids on all the containers and place them in the yoghurt maker, making sure that they are not touching the heat-sink or the resistor R4. Switch on, put the lid in place and leave for 5 to 7 hours. When the yoghurt is ready, it will have set to a stiff consistency. It may then be removed and, if necessary, stored for 3 or 4 days in a refrigerator. For making the next batch, you use the yoghurt that you set aside in the small container. If you make two batches a week in this way you will never need to buy another yoghurt. You will probably find that the second batch and subsequent batches (made with your own yoghurt culture) taste better than the first batch. If you want to make flavoured yoghurt you simple add chopped fruit, fruit-juice, or a spoonful of jam to the yoghurt and stir it well, just before you eat it.

Table 4.1. Components list for yoghurt maker

Resistors	
R1	thermistor, Siemens type K 164, or similar, 1kΩ at 25°C
R2	1kΩ carbon, ¼W
R3	22kΩ carbon, ¼W
R4	1Ω wire wound, 2.5W or, preferably 4W
Potentiometer	
VR1	Preset, horizontal, 4.7kΩ
Capacitor	
C1	220μF (higher value also suitable)
Semiconductors	
TR1	2N3704 (*npn*)
TR2	MJ491 (*pnp*)
D1	BYX38-300R
Transformer	
Mains primary, 9V secondary	
Miscellaneous	
DPST switch; stripboard, 0.15in matrix, 11 holes by 8 strips plus small scrap for mounting R4; two tins, such as biscuit tin and tea tin, 6 B.A. bolts, nuts and shake-proof washers (9 off); 4 B.A. bolts, nuts and shake-proof washers (2 off); 4 B.A. solder tag (1 off); sheet expanded polystyrene; grommets (3 off); sheet aluminium 18swg approx. 10cm X 6cm; connecting wire; solder; 3-core mains flex; mains plug	

5

Simple Touch Switch

This switch is triggered by touching a small plate with a finger. It gives effortless and silent operation and, as will be seen later, the touch plate can be made as conspicuous or as inconspicuous as you wish. It can operate small lamps, buzzers and motors powered by low voltage, or it can be used to switch any mains-powered device. This switch has a multitude of applications in the home, some of which will be suggested at various points in this chapter. The circuit is a simple one, employing only a handful of components, and its action too is simple. On touching the plate the lamp or other device is immediately switched on. It remains on for as long as contact is made with the plate, and for a period of about 10 seconds *after* the finger has been removed from the plate. This delayed switch-off leads to several useful applications, such as the operation of lamps for corridors the larder, cupboards, and porches. Such lamps are automatically extinguished after a period of 10 seconds unless contact is maintained. The switch can also be built to operate in reverse so that the device is normally on, but is switched *off* when contact is made.

The touch switch action depends on a field effect transistor (FET). As the symbol TR1 for the FET in Fig. 5.1 indicates, an FET can be thought of as a bar of conductive material. At one end is a contact, the source (s), through which electrons enter the bar when a potential is applied across its ends in the direction shown in the figure. At the other end of the bar is a second contact, the drain (d), through which electrons leave the bar (or drain away) and pass toward the positive terminal of the battery. The FET used in this circuit has its bar made from *n*-type semiconductor material. Along the sides of the bar are two regions of *p*-type semiconductor material, forming an electrode called the gate (g). When a potential is applied to the gate, an electric field is created in the *n*-type bar. The effect of this field is to restrict the width of bar available for the passage of electrons on their way from s to d.

The stronger the field, the more the passage of electrons is restricted. The action is rather like having water flowing along inside a soft plastics tube and then pinching the tube to restrict the flow. By varying the potential of g, we vary the strength of the field, and thus control the

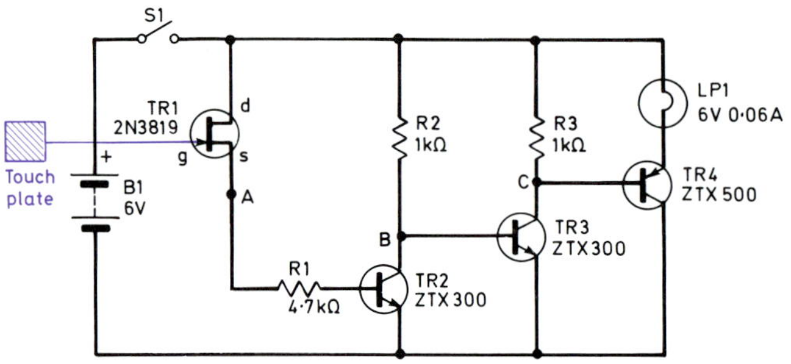

Figure 5.1

Circuit diagram of the simple touch switch; the lamp lights when the touch plate is contacted

flow of electrons from s to d. The current passing along the bar is varied, even though the potential difference between d and s remains constant. In other words, we vary the *effective resistance* of the bar – we are using the FET as a variable resistor. Instead of turning a spindle to vary the resistance, as we do with an ordinary volume-control type of variable resistor, we simply alter the potential at g. An FET can also be used as an amplifier, but in this circuit we are not concerned with this kind of application.

The circuit

One important property of the FET is that the input impedance of its gate electrode is extremely high. When the potential at g is varied, the current entering or leaving g may be only a few picoamperes ($1\text{pA} = 10^{-12}\text{A}$), yet the resulting field effect causes appreciable changes in the effective electrical resistance between s and d. This property is fully exploited in the touch-switch circuit. When we touch a finger against the touch plate, which is connected to g, the resulting change of potential at g causes a sudden increase in the resistance between s and d. The resistance changes from a few hundred ohms to several kilohms, with the result that the potential at point A (Fig. 5.1) suddenly falls from almost 6V to only a fraction of a volt above zero. We make use of this large change of potential to operate the switching transistors in the remainder of the circuit.

If the potential at A falls, the base current that was formerly being supplied to TR2 through R1 is suddenly reduced almost to zero. TR2 is turned off. This causes the potential at its collector (point B) to rise sharply to almost 6V. This turns on TR3, making the potential at C fall very low, as TR3 begins to conduct. The fall in potential at C turns TR4 *on*, for TR4 is a *pnp* transistor. Collector current lights lamp LP1, and the lamp remains on for as long as the finger is touching the plate. After contact is broken, the potential at A rises slowly back to almost 6V; then B goes low, C goes high and TR4 is turned off, gradually extinguishing the lamp.

A circuit such as this could be used to illuminate a cupboard or porch, giving you time enough to find what you want in the cupboard, or to select the correct door key and insert it in the lock. Since the whole circuit is battery powered it is unnecessary to become involved with mains wiring. The circuit requires only 4mA when in its inactive or quiescent state. It can be switched on (by closing S1) at darkness each day, and switched off at bedtime. With such usage the battery will have a reasonably long life. If it is to be left in operation for longer periods, or continuously, it is more economical to build a mains power supply. The one used to power the light operated switch (Project 9) is suitable for this purpose.

Although Fig. 5.1 shows a lamp, this circuit can also be used for switching other devices that operate on a 6V supply. Instead of LP1, you can connect the oscillator circuit of the moisture detector; omit B1 and S1 of Fig. 1.1, and also the probe and its leads; connect the positive ends of R1 and R2 to the positive rail of the touch-switch circuit; connect the negative rail of the oscillator circuit to the collector of TR4. A warning sound is produced whenever the touch plate is contacted and sounds for about 10 seconds afterward. It could have various warning applications, including that of an intruder alarm. Touch plates concealed in strategic places could soon reveal the presence of a person not 'in the know'. The system can also be used as a door alarm, though the curious wailing sound it makes after switch-off renders it less desirable in this application. It is also possible to connect an ordinary door buzzer or bell in place of LP1, but if this is done a diode *must* be connected across its terminals, as shown in Fig. 5.3. (The reason for this is explained later.) Similarly, the circuit can switch a small motor, but a diode must be included across its terminals; this could have several applications in the operation of working models. One point to remember is that the current passed by the switched device must not exceed the current rating of TR4. For the ZTX500 the maximum current is 0.5A. For switching larger currents, use a *pnp* power transistor such as MJ491, which can handle up to 4A. If the power is to be switched on for prolonged periods, it will be necessary to mount the transistor on a heatsink.

Before proceeding to constructional details we should consider the possibilities of two other variations on this circuit, both being slightly simpler than the version of Fig. 5.1. In Fig. 5.2 we have a circuit which operates with the reverse action; it switches *off* the device (lamp, buzzer, motor) for as long as the touch plate is being contacted. In Fig. 5.3 we see a circuit designed to operate a relay. A relay would

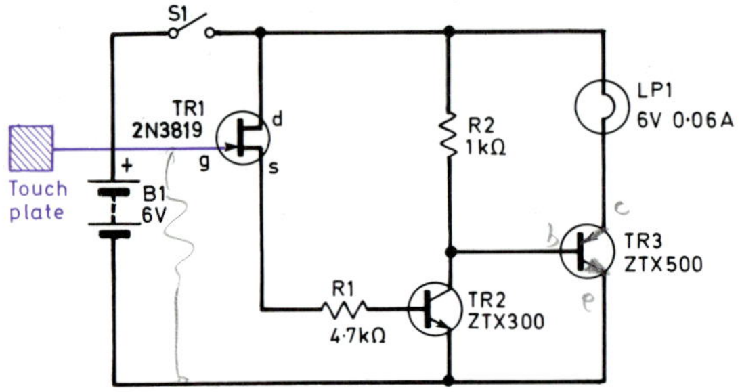

Figure 5.2

Alternative circuit for a touch switch, in which the lamp is normally on and extinguished when the touch-plate is contacted

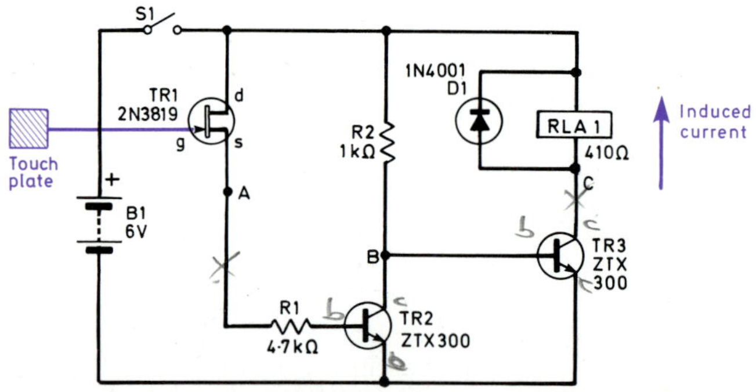

Figure 5.3

A simple touch switch circuit for operating an electromagnetic relay

work in place of LP1 in the circuit of Fig. 5.1, but since the relay coil has a relatively high resistance it can take the place of R3 in the collector circuit of TR3, and we can dispense with TR4. When the plate is touched the potential at point C goes low, current flows through the relay coil and its armature is actuated. Depending on the construction of the relay, one or more contacts may be made or broken simultaneously. We can perform many and complex switching functions in this way, though the relay recommended for this circuit operates simply as a single-pole double-throw switch, enabling you simply to switch a device on, or off (or one device on and another off), for as long as the plate is touched. The switch-off action is sharp, so the relay could well be used for controlling the oscillator of Project 1, as a simple door-alarm; for a pleasanter sound you could use the door alert of Project 8. Since the relay can switch mains current, it has a wide range of possible applications for switching room lights, motors, heaters and other mains-powered devices. The use of the relay for switching mains power is explained in more detail in Project 9.

A relay, like an electric motor and an electric bell, is an *inductive* device, in which one or more electrical coils are used to produce a magnetic field. When the current to such a device is cut off, the magnetic field collapses instantly. The effect of this rapid collapse is to *induce* a potential between one terminal of the coil and the other terminal. As a result of this potential a current flows through the coil *and through the circuit to which it is connected.* The direction of this induced current is always *opposite* to that of the supply current which was previously cut off. In Fig. 5.3, the direction of the induced current is indicated by the arrow. The induced potential and the current it causes are very short-lived, but their magnitude is high. The potential can easily exceed 100V in a circuit such as this, and may be greater in other circuits. Indeed, it is this principle which is used in the magneto of a car engine to produce the high potential required for the sparking plugs from the low potential between the terminals of the car battery. Such high voltages can cause instant damage to a transistor, especially as their direction is the reverse of the normal supply voltage. This is the reason for connecting D1 across the terminals of RLA1. At switch-off, the high induced potential is immediately discharged through D1, and does not affect the remainder of the circuit.

Construction

Construction of any one of the variations of the simple touch switch presents few problems. The layout of the components on the board is not critical; Fig. 5.4 shows a suggested arrangement for the circuit of Fig. 5.3. If built on a piece of stripboard of the size shown, the circuit can be housed in a plastics case like that recommended for Project 9 (Fig. 9.5). The board is slotted into guides moulded in the walls of the case. A second board of identical size carries the mains power unit (Fig. 9.4) if this is required for continuous operation of the touch switch circuit.

First solder TR1, R1 and R2 into position, and also the three terminal pins. Connect the pins at A3 and O3 to a battery and connect a voltmeter between the negative rail (pin O3) and the point corresponding to A in Fig. 5.3 (the end of R1 that connects with the source of TR1).

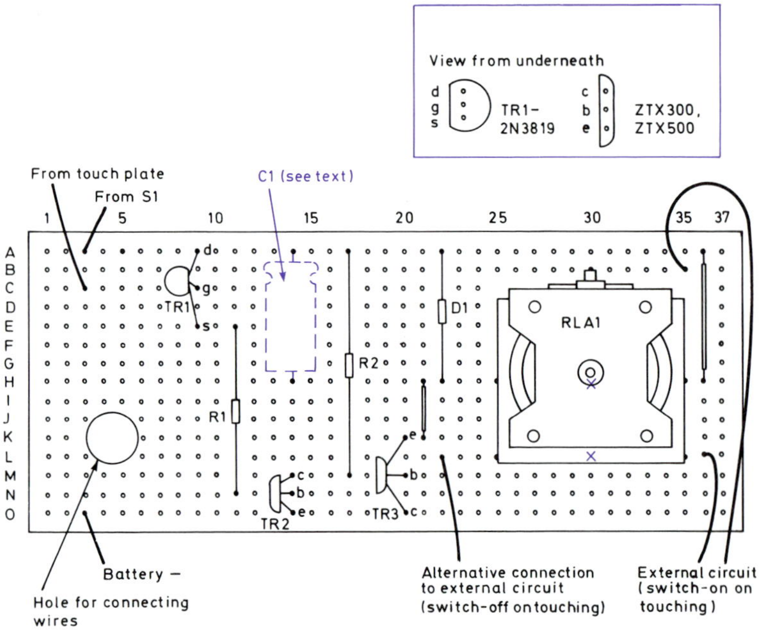

Figure 5.4

Layout of the circuit board for the touch switch circuit of Figure 5.3

The voltmeter should read almost 6V. Touch your finger briefly against the terminal pin at C3. The reading should fall sharply, almost to zero, then slowly rise to 5V. If you hold your finger against the pin, the reading should stay low until you remove the finger. You have now demonstrated the triggering action of the touch switch and can next experiment with the construction of a suitable touch plate. There are many ways of making a touch plate and the method you choose will depend on the application you have in mind. The essential point is that the plate must be metal and that it must be adequately insulated from earth and from the remainder of the circuit. If the circuit is to be housed in a plastics case, the plate can be mounted on the outer wall of the case. It can be a square or disc of metal foil. Self-adhesive metal foil can be used conveniently, or you can attach a piece of kitchen foil using contact adhesive. You can make the touch plate large, so that it can easily be found in the dark (or by blind persons). You will find that in practice it is not necessary for the finger actually to make contact with the plate. A thin layer of plastics film may be placed over the plate, yet the switch will still operate when the film is touched. Here the capacitative action between finger and plate, with the plastics film as dielectric, is sufficient to affect the FET. This means that you can cover your touch plate with polythene sheet or with decorative plastics sheet such as *Contact.* The plate can thus be held in position and be protected from corrosion. It can also be made less conspicuous if you wish. This method could be used for mounting the plate on a wall or door. The plate can also be concealed beneath wall-paper, provided the wall is a dry one. A simple way of making a touch plate (this necessarily being a small one) is to push an ordinary drawing-pin through the plastics wall of the case. The lead to the terminal pin is soldered to the point of the pin. Take care when inserting the pin, for some kinds of plastics crack when a thick pin is pushed into them. It is better to drill a hole first – just a little narrower than the diameter of the pin – so it will grip the pin firmly when in place. As already explained, finger-to-metal contact is not essential; you can use a pin with a plastics-covered head. If several touch switches are contained in one case, use differently coloured pins for each touch plate.

Having decided on the type of touch plate to use, and having tested it on the partly built circuit to check that it really triggers the circuit, you can proceed to complete the construction of the circuit. The remaining components should be soldered into position, as shown in Fig. 5.4. If you are building the circuit of Fig. 5.1, you will be able to find plenty of room at the right-hand end of the board to accommodate TR4 and R3. If you are intending to use the relay to switch mains current, an additional component is required. This is a capacitor C1 (100μF or more) which is placed where indicated. It has been found that without the capacitor there are electrical disturbances (presumably due to sparking as the relay contacts open) which continually trigger the FET and prevent the circuit from switching off when once it has been switched on. A capacitor connected in parallel with the relay coil eliminates this trouble. If you find that the relay 'jitters' at switch-off, substitute a capacitor of higher value, say 470μF or 1000μF. Points about wiring the relay for mains operation are dealt with in Project 9. Read the appropriate section before proceeding further, and remember that, with the type of relay recommended, parts of the framework of the relay are live – handle with great care when testing and, house the project in a case of plastics or other insulating material. The finished appearance of the project inside its case is very similar to that of Project 9 (Fig. 9.7).

Table 5.1. Components list for touch switch

	Fig. 5.1	*Fig. 5.2*	*Fig. 5.3*
Resistors			
R1	4.7kΩ	4.7kΩ	4.7kΩ
R2	1kΩ	1kΩ	1kΩ
R3	1kΩ	—	—
all carbon, $^1/_4$W			
Capacitor			
C1	—	—	100μF, 10V (for mains switching only, see text)
Semiconductors			
TR1	2N3819	2N3819	2N3819
TR2	ZTX300	ZTX300	ZTX300
TR3	ZTX300	ZTX500	ZTX300
TR4	ZTX500	—	—
D1	(1N4004 required if inductive load to be switched, see text)	—	1N4001 obligatory
Miscellaneous			
LP1	6V 0.06A (or other switched devices, see text)	6V 0.06A	—
RLA1	—	—	Miniature open printed-circuit relay, 6V, 410Ω coil

Stripboard, 0.1in matrix, 15 strips by 37 holes; terminal pins (6 off); materials for constructing touch-plate (see text), plastics case (for example a.b.s. case, type M3); connecting wire; solder. Components required for construction of mains power supply circuit are listed on p. 57.

6

Latching Touch Switch

Although the short delay action of the touch switch described in Project 5 has several useful applications, it does not give the fully controlled switching action that is required for many purposes. For example, you may wish to switch on the lamp of a photographic enlarger for a precisely determined period. Touch control is very useful for such a switch, for the absence of vibration and the ease of finding the touch plate in the dark are both advantages with an enlarger switch. The switch described in this project can be turned off or on exactly when required. It makes use of two FETs, and two touch plates which would normally be mounted side by side. You touch one plate to turn the switch on, and you touch the other plate to turn it off again.

The circuit

The switch depends for its action on a bistable circuit. This is an arrangement of two transistors that can remain stable in one of two states (Fig. 6.1). In state A, TR1 is off, and because it is not conducting its collector is at high potential (point A). This allows base current (dashed line) to flow to TR2, through R1 and R4, turning it on. A collector current flows through TR2 (continuous line), so its collector voltage is low (point B). This means that the current flowing from B through R2 to TR1 is insufficient to turn TR1 on. The circuit is stable in this state. It can also remain stable in state B, in which TR1 is on and TR2 is off. Since the circuit has two stable states, we call it a *bistable circuit.*

The circuit can be made to change from one state to the other by momentarily removing or reducing the base current to the conducting transistor. If it is in state A, and we momentarily connect point D to the negative rail, TR2 is deprived of base current, it immediately ceases to conduct, the potential at its collector (point B) rises, supplying base current to TR1, which then switches on. This causes the potential at A to fall low, and the bistable is now in state B. Even if we now *remove* the connection between D and the ground rail, insufficient

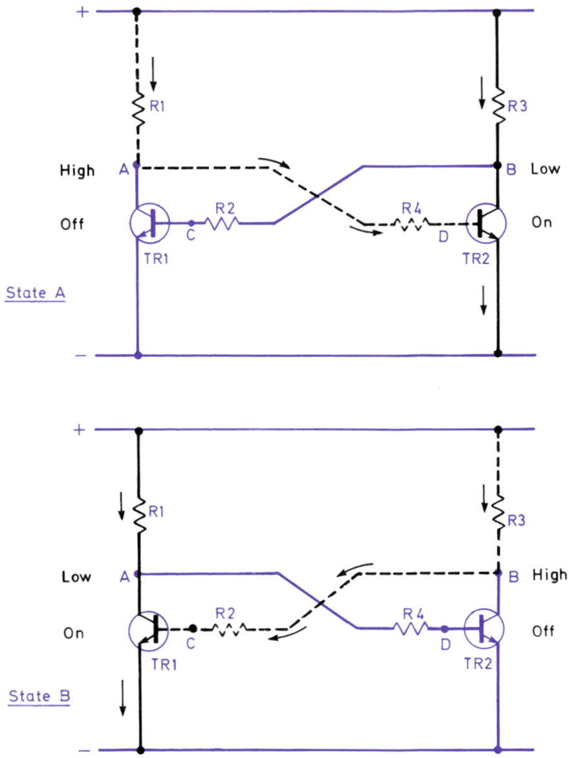

Figure 6.1

The action of a bistable circuit

base current can flow to TR2, for the potential at A is low. To make the circuit change back to state A again, we momentarily connect C to the ground rail. Thus we have a circuit which can be made to change state by either grounding a transistor base terminal, or by applying a short negative pulse to a base terminal. In the touch switch the negative pulses are produced by the FETs (Fig. 6.2). When the *on* plate is touched, the potential at point A falls sharply, as explained on p. 26. This draws current from C1 through D1, causing a brief negative-going pulse at the opposing plate of C1. This is sufficient to reduce the base current to TR3 below the level required to keep it switched on. If it was previously in state B the bistable changes to state A. If it was previously in state A, it remains in that state. Similarly, touching the *off* touch plate causes the bistable to change to state B, if not already in that state. Note

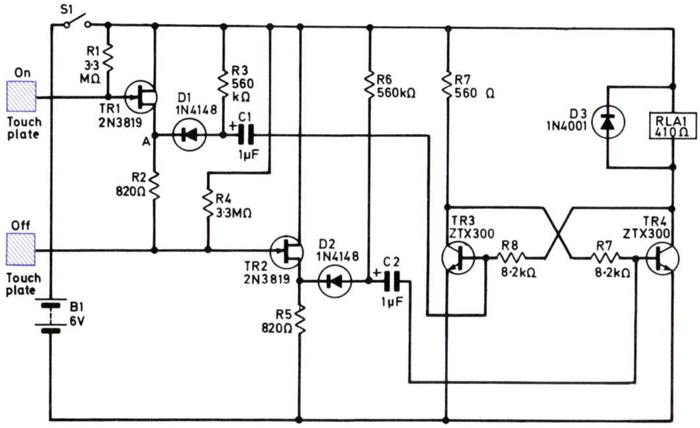

Figure 6.2

Circuit diagram of the latching touch switch

that instead of a resistor in the collector circuit of TR4 we have the coil of RLA1. The coil is energised when the bistable is in state A (after touching the *on* plate) and de-energised when the bistable is in state B (after touching the *off* plate). It should be mentioned that in order to produce a sharp triggering pulse from a plastics-covered touch plate it is necessary to *tap* the touch plate lightly rather than merely touch it. In general, a bare metal plate is better for this project. It will be recalled that in the simple touch switch (Project 5) the potential at A rises fairly slowly after its initial sharp fall: this inactivates the *on* switch for about 10 seconds after it has been operated. This produces a delay which is made good use of in Project 5, but in this project it would be impossible to turn the relay on and off in rapid succession. This difficulty is overcome by the resistors R1 and R4, which allow a tiny current to flow to the touch plates and bring them rapidly back to their original potential. Since we are triggering a very sensitive circuit – the bistable – which can respond to the very brief initial drop of potential, the addition of these resistors does not upset the switching action of the FETs.

The circuit can be set out on a rectangle of stripboard as in Fig. 6.3. The layout bears some similarities to that of Project 5 and, like the board of that project, can be slotted into the kind of case recommended

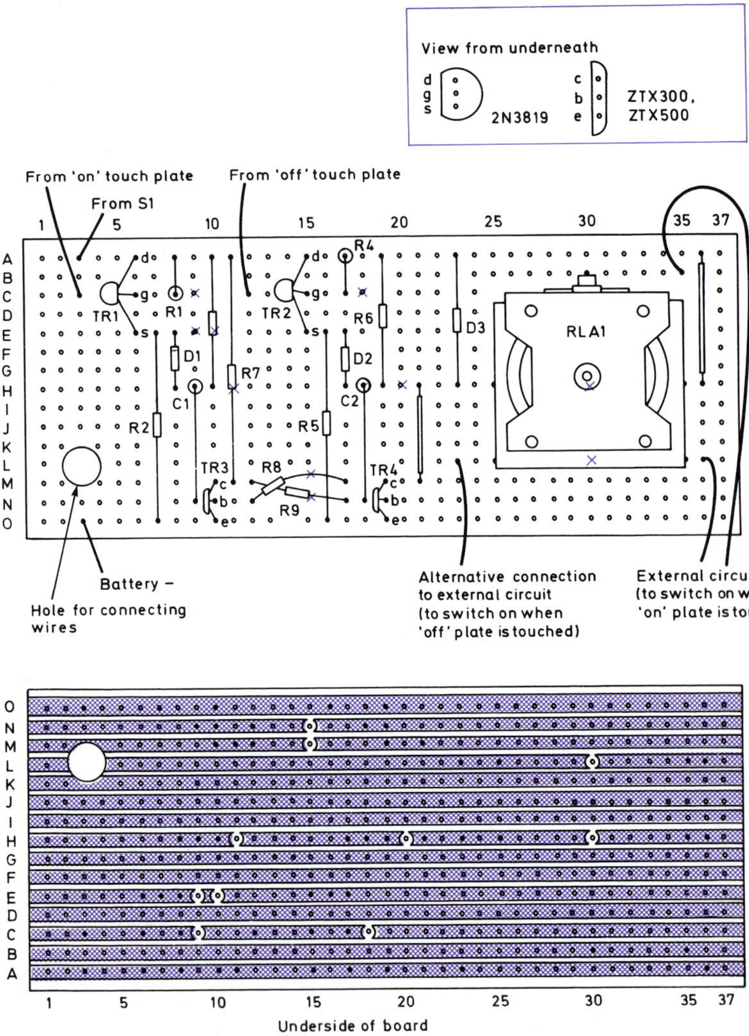

Figure 6.3

Layout of the circuit board for the latching touch switch

for Project 9 (Fig. 9.5). Although this project *may* be powered from a 6V battery (its current requirement is only 17mA when *off* and 20mA when *on*), it is likely that a constructor who has gone to the trouble of building this more elaborate switch will most likely wish to build a mains powered supply for it. The circuit used in Project 9 (Fig. 9.4) is very suitable for this purpose. The latching touch switch circuit is not affected if the relay is used to switch mains currents, so the additional capacitor used in Project 5 is *not* required in this project.

Construction

First complete the bistable by soldering TR3, TR4, R7–9, D3 and RLA1 in place. To test this, connect a 6V battery to A3 and O3. Connect a length of wire to the negative terminal of the battery and touch this alternately against the collector wires (c) of TR3 and TR4. When TR3 collector is grounded the relay should click, with its coil energised (*on*, state A). Further contact at the collector of TR3 has no effect. But if the wire is next touched to the collector wire of TR4, there is a second click as the relay coil is de-energised (*off*, state B).

Next make up the two touch switches with touch plates (see Project 5 for ideas for touch plates, but remember that *bare* metal is preferred for this circuit). For the moment, do not solder capacitors in position. The touch switches may be tested to check that the potential at the anodes of D1 and D2 (at H8 and H17) fall sharply when the touch plates are contacted. If all appears to be in working order, the capacitors may now be soldered in place to link the touch switches to the bistable. The circuit should then respond rapidly to the lightest contact with the touch plates. It is possible to repeat the

Figure 6.4

Alternative circuit for switching a low-voltage load instead of a relay

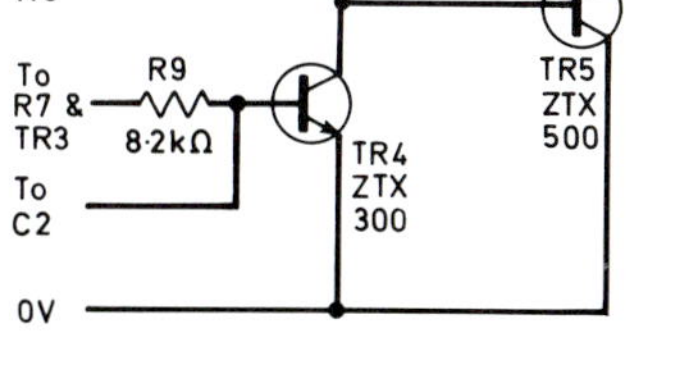

on-off sequence many times per second, and the circuit faithfully responds. This provides a reliable and rapid-acting switch suitable for switching room lamps, warning lights, enlarger lamps and photographic floodlights, as well as for controlling the power to motors, radio and TV sets, and for almost any other switching function. As with Project 5, it is possible to dispense with the relay if you are controlling only low-voltage equipment. If you wish to do this, TR4 can be made to

Table 6.1. Components list for the latching touch switch

Resistors	
R1	3.3MΩ
R2	820Ω
R3	560kΩ
R4	3.3MΩ
R5	820Ω
R6	560kΩ
R7	560Ω
R8	8.2kΩ
R9	8.2kΩ
all carbon, ¼W	
Capacitors	
C1, C2	1μF tantalum (two off)
Semiconductors	
TR1, TR2	2N3819 (two off)
TR3, TR4	ZTX300 (two off)
D1, D2	1N4148 (two off)
D3	1N4001
Miscellaneous	
RLA1 Miniature open printed-circuit relay, 6V, 410Ω coil (Doram Electronics Ltd)	
Stripboard, 0.1in matrix, 15 strips by 37 holes; terminal pins (six off); materials for constructing touch plates (see text), plastics case (for example, a.b.s. case type M3); connecting wire; solder. (Components for construction of mains power unit are listed on p. 57).	

operate a switching transistor TR5 instead of RLA1. The required circuit modification is shown in Fig. 6.4; the additional components can readily be accommodated in the board space occupied by RLA1 in Fig. 6.3. If the load is non-inductive (lamp or UJT oscillator circuit), D3 may be omitted.

7

Delay Switch

This has a function similar to that of the simple touch switch of Project 5, but performs with greater reliability. The delay period of this switch is adjustable to suit your requirements and can be made quite lengthy. For example, you can switch on your bedside lamp or radio set and know that it will be switched off automatically 30 minutes later, even if you have fallen asleep in the meantime. As a switch for a hall or landing light, you can set it to remain on long enough to give even the slowest-moving member of the family ample time to climb up or down the stairs before it automatically switches off. For these applications the switch is constructed to operate for a single pre-set period but, if you wish, you can arrange for the period to be variable, with a calibrated control knob. This makes a useful switch for an enlarger lamp, enabling you to set the exact length of exposure required. Timing can be set very accurately, giving reliable exposure control. If preferred, the device may be triggered by an ordinary pushbutton instead of a touch plate; this allows the possibility of triggering it remotely.

The circuit of the 555

The circuit is entirely different from that of the previous switching projects. It is based on a single integrated circuit, a most useful and versatile device known as the 555 timer i.c. This contains almost all the components needed to make either an astable multivibrator (we use it for this purpose in Project 11) or a monostable multivibrator, as in this project. A monostable multivibrator is a circuit that can exist in either of two states. When triggered it goes to its second state and remains in this state for a period of time, then automatically reverts to its original state, where it remains until it is again triggered. The timer is shown in Fig. 7.1, connected to make a monostable multivibrator. It is connected to the power supply through pins 1 and 8; the supply voltage (V_{CC}) may be anything between 4.5V and 16V. In this project V_{CC} is approximately 6V, provided either by a battery or the power unit of Project 9 (Fig. 9.4). The trigger input (pin 2) may be connected to a touch plate of the kind described in Project 5.

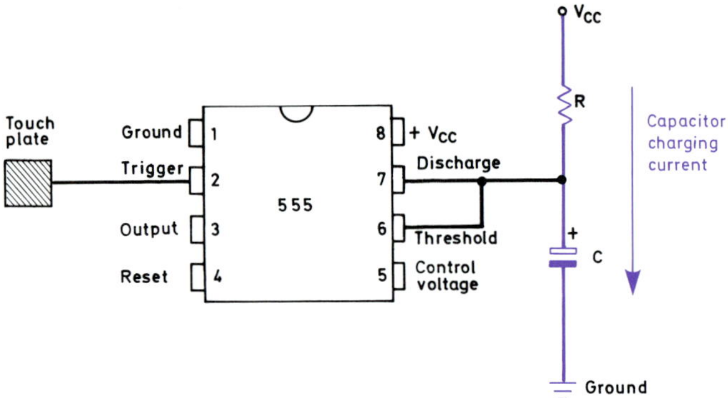

Figure 7.1

The 555 integrated circuit connected as a monostable multivibrator

This input is very sensitive and is often triggered simply by the hand being brought toward the touch plate. This extreme sensitivity can be a valuable feature, or it can be a nuisance. If the layout of your delay switch turns out to be prone to spurious triggering and you find this a nuisance, you may trigger the 555 by using a pushbutton switch to momentarily connect pin 2 to ground. When triggered, the output (pin 3) rises from low (close to 0V) to high (almost to V_{CC}). A maximum current of 200mA may be drawn from the output, so the i.c. can be used to drive certain devices (other i.c.s, LEDs etc) directly. In this project we use the output to supply base current to a switching transistor. Before the i.c. is triggered the potential at pin 7 is $\frac{1}{3}$ of V_{CC}, (about 2V in this project). Current flowing through R from the positive rail flows in at pin 7, the potential across C remaining steady at $\frac{1}{3}$ V_{CC}. As soon as the i.c. is triggered the circuit immediately internal to pin 7 becomes non-conducting. Consequently the potential across C gradually rises. The rate at which it rises depends on the capacitance of C and the resistance of R. The potential rises until it reaches $\frac{2}{3}$ V_{CC} (about 4V); when this threshold voltage is reached it is detected by the circuit internal to pin 6. This triggers the output to fall to its low condition, and simultaneously C is discharged through pin 7 to $\frac{1}{3}$ V_{CC}.

Thus, by choosing suitable values for C and R, we can arrange for the output to go high for a predetermined period after triggering. This period can be of the order of microseconds or as long as an hour or more.

The reset input (pin 4) is used to make the output return to its low condition before the end of the set time period. Normally pin 4 is connected to V_{CC}; to reset the timer pin 4 is momentarily connected to ground. The control voltage input (pin 5) is used to alter the threshold voltage at which the i.c. operates; we do not use this facility in this project.

The circuit of the delay switch

The practical circuit is shown in Fig. 7.2. The output pin is connected to TR1 by R1, through which a base current flows when output goes

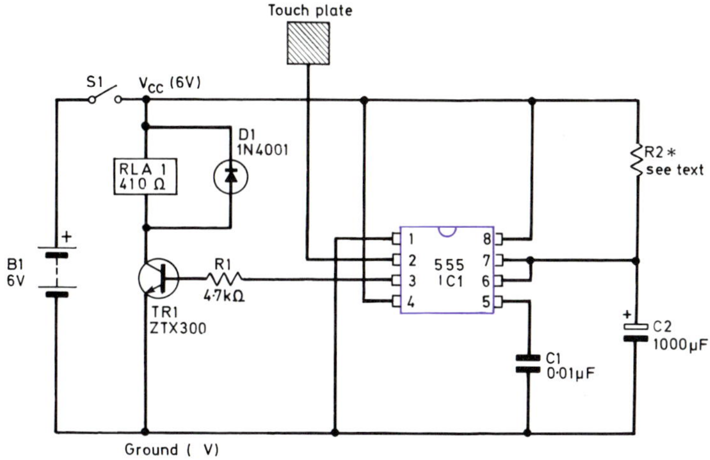

Figure 7.2

Circuit diagram of the delay switch

high. This turns on TR1 so that the potential at its collector goes low, and the relay coil is energised. Other inductive loads may be substituted for RLA1, or a non-inductive load such as a 6V, 0.06A lamp may be used, in which event D1 is not required. The period of time for which the switch operates is determined by the value chosen for R2. In the prototype it was found that when R2 was 10kΩ the switch operated for 15 seconds; when R3 was 100kΩ it operated for 3 minutes; when R2 was 1MΩ, the period was 30 minutes. Using these figures as a rough guide, the reader will be able to choose a suitable value for R2. It is, of course, possible to use a capacitor with capacitance higher or lower than 1000μF; the higher the value, the longer the period of operation. For really long periods of operation C2 could be as high as 10,000μF and R2 as high as 20MΩ, but leakage of current through C2 would account for much of the current passing through R2 and pin 6 might never reach the threshold potential. A period of an hour or so is about

Figure 7.3

Modifying the circuit of Figure 7.2 for operation by push-button switches

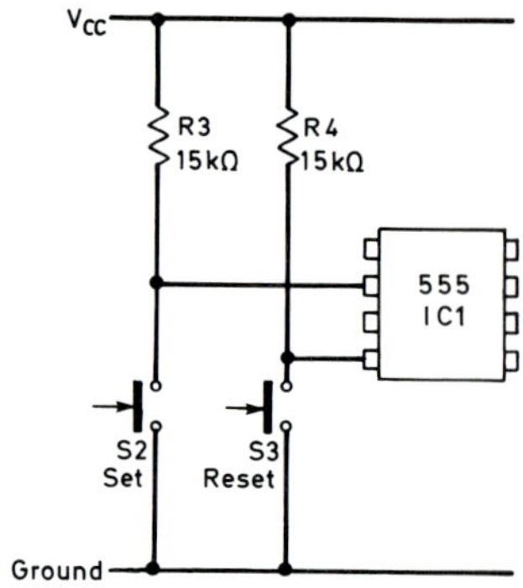

as long as can be reliably attained in practice with this circuit. A modification of the circuit for control by pushbuttons is shown in Fig. 7.3.

For many applications the exact length of delay is not important. R2 can be selected from the range of preferred values. If a period of exactly specified length is required, R2 may be replaced by a fixed resistor and a preset variable resistor wired in series. The variable resistor is then adjusted until the switch operates for the required period. To make a timing switch for use with a photographic enlarger, R2 is replaced by a potentiometer of suitable value (say, 4.7kΩ).

Construction

The components may be placed on the circuit board as shown in Fig. 7.4. This size of board is chosen to fit in the type of case shown in Fig. 10.5, which can also contain the power unit described in that project. The circuit does not really need as much board space as shown in Fig. 7.4, and can easily be accommodated on a smaller board to allow it to fit in a smaller case, if desired. If pushbutton operation is to be employed, the left-hand end of the board is set out as shown in Fig. 7.5. The other terminals of S2 and S3 are grounded, either to the pin at O3, or to the negative terminal of the power supply.

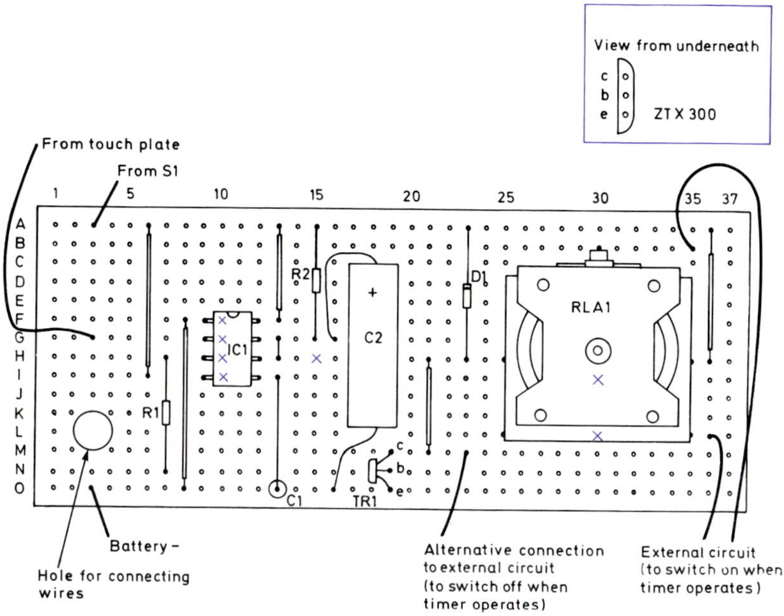

Figure 7.4

Layout of the circuit board for the delay switch

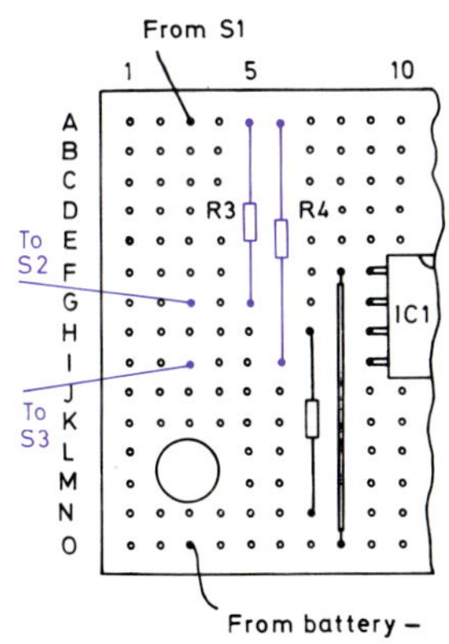

Figure 7.5

Modifying the layout of Figure 7.4 when push-button switches are used

Whether or not to use an i.c. socket is a matter of personal preference. Beginners may feel safer when using a socket, and it does make it easier to remove the i.c. when testing for faults. But the circuit is a simple one, with little to go wrong so anyone with a small amount of experience need fear no trouble if a socket (relatively expensive) is dispensed with and the i.c. soldered directly to the board. First solder in IC1 and the connecting wires to it. Do not forget the solder blob below the board to connect pin 6 to pin 7. Then solder in R2, C2 and C1. It is advisable not to trim the wires of R2 yet, as you may need to substitute another resistor after testing. Temporarily solder a piece of connecting wire in hole H7 and connect this to the positive terminal of a voltmeter; connect the other terminal to the ground rail (pin O3). Connect a 6V battery or the power unit to the board at A3 and O3 and switch on. Touch the touch plate; if you do not have the plate ready yet, the i.c. should easily be triggered merely by touching the pin at G3. The output as read on the voltmeter should go to 6V, remain steady at 6V for the required period, and then go low. The rise and fall of output voltage are both rapid. If the circuit fails to operate as described, check the connections. When checking, you can speed the operation of the timer by temporarily connecting an additional resistor (say 1kΩ) in parallel with R2. When all appears to be working (though with a shorter time period than you eventually require), remove the additional resistor and see if the timer now operates for the correct period. If the period is too short, replace R2 with a resistor of higher value; if the period is too long, replace R2 with a resistor of lower value. It is worth remembering that the capacitance of an electrolytic capacitor is not necessarily close to the nominal value marked on its case. In particular an old capacitor or one that has not been in use for several months may have a reduced capacitance, and a relatively high leakage current. This will affect the timing considerably. Unless you are using a capacitor newly purchased from a supplier known to maintain fresh stocks it is worth while to connect the capacitor you intend to use across the terminals of a 6V battery or the 6V power unit for an hour or so before you build it into this circuit. In that time the electrolytic dielectric will have had time to re-form and the capacitor will regain its lost capacitance and reach a more-or-less stable value.

When the timing period has been checked and found to be correct, complete the construction of the circuit by soldering R1, TR1, D1 and RLA1 in place. There should be no problems in getting the relay to trip as soon as pin G3 is touched and to remain in that state for the desired period. However, under some conditions you may find that the i.c. is re-triggered as it switches off; this may be caused by mains-borne spikes, perhaps originating from the relay or from the device being switched by the relay. In this instance the extreme sensitivity of the i.c. trigger input is a disadvantage. In these circumstances it is best not to use a touch plate, but to use a pushbutton. If you think that you may wish to switch the device off before the timer has run its full period you may also fit a pushbutton to the reset. This is optional and is hardly worthwhile if the timer is to be used for periods of only a few tens of seconds.

Table 7.1. Components list for delay switch

Resistors	
R1	4.7kΩ
R2	value depends on period required (see text)
R3, R4	15kΩ (required only for pushbutton version)
all resistors carbon, $^1/_4$W	
Potentiometer	
Preset or panel-mounting, required only for adjustable versions	
Capacitors	
C1	0.01μF, ceramic disc
C2	1000μF electrolytic, 6V or 10V
Semiconductors	
TR1	ZTX300
D1	1N4001
Integrated circuit	
IC1	NE555, timer i.c.
Miscellaneous	
RLA1 Miniature open printed-circuit relay, 6V, 410Ω coil, stripboard, 0.1in matrix, 15 strips by 37 holes; terminal pins (6 off); materials for constructing touch plate, if used, *or* pushbutton switches (1 or 2 off, as required); plastics case (for example, a.b.s. case type M3); connecting wire; solder (Components for construction of mains power unit are listed on p. 57).	

For accurate setting of the time period there is room on the board for a preset resistor to be connected *in series* with R2. If you are building a calibrated timer, omit R2 and solder a terminal pin at hole G15. From this pin run a connecting wire to the wiper of a potentiometer, mounted on the panel of the instrument case. Run a wire from one end terminal of the potentiometer to the pin at A3 (V_{CC}).

8

Door Alert

The original name for this project was Door Alarm, but when the prototype was completed it was found that its sound was far from alarming. Instead, it gives a pleasantly musical chord, rather like that

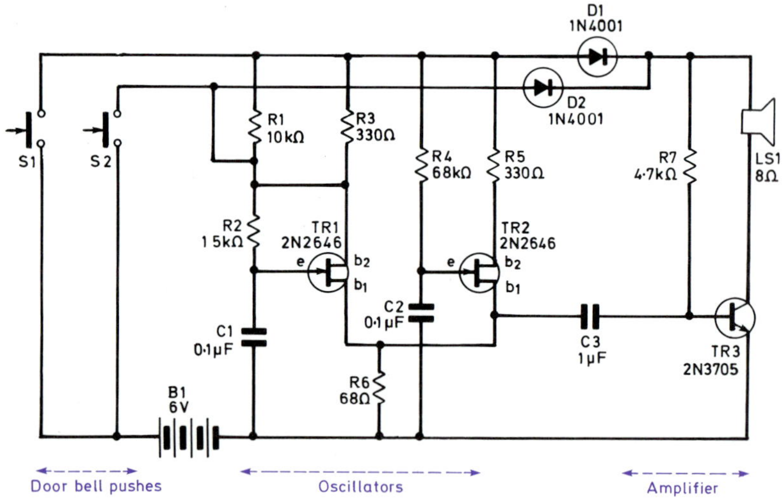

Figure 8.1

Circuit diagram of the door alert

emitted by the simpler types of electronic organ. Hence the change of name to Door Alert. I must confess that the final *design* – if that is truly the correct word – owes something to a lucky chance, including an accidental short circuit on the patchboard, which lead to a rather unusual way of joining together two unijunction relaxation oscillators (Fig. 8.1). Comparison with Fig. 1.1 shows that there are two such oscillators here, but the connections between them are complex and to a certain extent fortuitous. It would be hard to say *exactly* how the coupled oscillators produce their complicated waveform, so let us rest content with the fact that they do, and that they give the effect of a musical chord when button S1 is pressed, and a different chord when S2 is pressed. The two chords differ in pitch and quality, so that anyone except the most tone-deaf can distinguish between them. This gives us the basis of a door alert system that can be sounded from either of two doors, and you can tell at which door the caller is waiting by listening to the chord produced.

The circuit

The signal produced by the oscillators is transmitted through C3 to the amplifying transistor TR3. The variations in base current caused by the signal result in corresponding variations in the collector current. This passes through the loudspeaker LS1 producing the alerting sound. Note that power is supplied to the circuit only when one or the other of the buttons is pressed. At other times no current flows, so this circuit is ideal for battery operation.

Construction

Fig. 8.2 shows how the components may be assembled on a piece of stripboard. First assemble the amplifier by mounting R7, C3 and TR3, and connecting LS1 temporarily to the pins at A25 and C25. Connect a 6V battery across pins A1 and I1. Connect a loose wire to the lower end of C3 (at hole H21). Touch this first against the positive battery terminal and then against the negative terminal; loud clicks from the speaker indicate that the amplifier is probably working correctly.

Next assemble the oscillators. If you have a patchboard (such as an S-Dec or T-Dec) you can set out the oscillators on this first, and experiment with various values of resistor for R1, R2 and R4 to see what unusual sounds you can obtain. The patchboard oscillators can be temporarily connected to C3 so that you can hear the weird musical effects you generate. Having decided on values that give a pleasantly melodious pair of tones, or having decided that you prefer the values recommended in Fig. 8.2, complete the soldering of components to the stripboard. With the battery connected to I1 and either A1 or D1 the appropriate chord should be heard loudly.

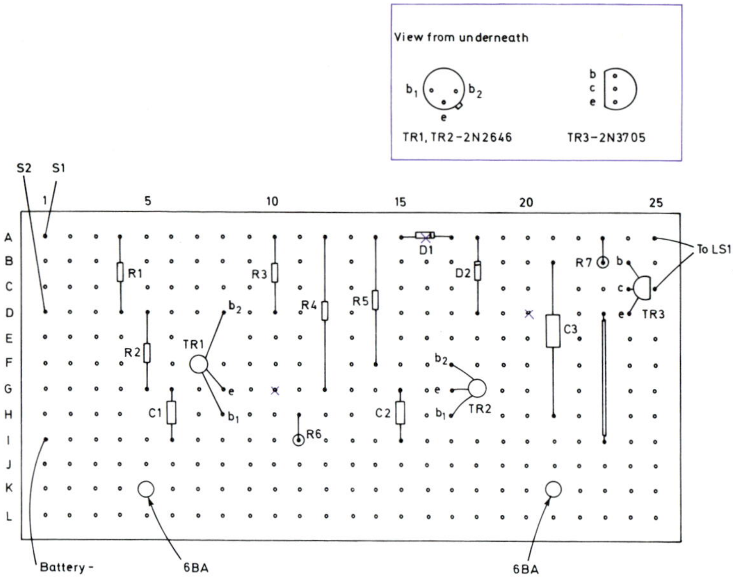

Figure 8.2

Layout of the circuit board for the door alert

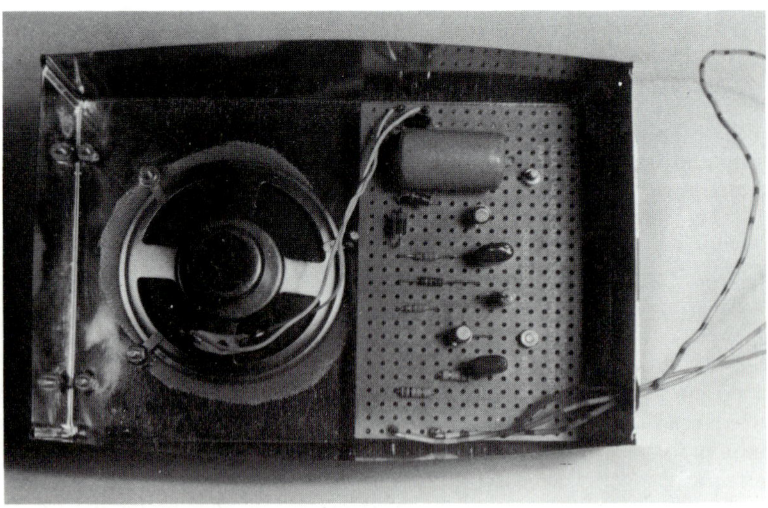

Figure 8.3

Rear view of the door alert, with lid removed showing the loudspeaker and circuit board in position

The circuit can be contained in a metal or plastics case, similar to that shown in Fig. 8.3. A large circular hole is cut in the front of the case and the loudspeaker is mounted behind this. Three holes drilled at points equally spaced around the large hole have a bolt threaded through them, and a nut tightened on the bolt. The speaker is then placed between the bolts. A solder tag is slipped on each bolt and held in place by a shakeproof washer and second nut. As the nut is tightened, the end of each tag is turned inward to grip the rim of the loudspeaker frame. A piece of stiff fabric or wire mesh may be placed behind the large hole and held in place by the speaker. Alternatively,

Table 8.1. Components list for the door alert

Resistors	
R1	10kΩ
R2	15kΩ
R3	330Ω
R4	68kΩ
R5	330Ω
R6	68Ω
R7	4.7kΩ
all carbon, ¼ W	
Capacitors	
C1, C2	0.1μF polyester (two off)
C3	1μF polyester
Semiconductors	
TR1, TR2	2N2646 unijunction transistors
TR3	2N3705
D1, D2	1N4001 (two off)
Miscellaneous	
Stripboard 0.15in matrix, 12 strips by 25 holes; terminal pins, for 0.15in matrix board (five off); loudspeaker, 8Ω; case; material for loudspeaker grille; grommet; push-switches (two off) and fixing screws; 6V battery with battery connectors; 6 B.A. bolts (five off); 6 B.A. nuts (10 off); 6 B.A. shakeproof washers (five off); 6 B.A. solder tags (three off); solder; connecting wire	

a loudspeaker grille may be bolted on to the front of the case, covering the loudspeaker hole. It is sometimes possible to recover a suitable grille from an old transistor radio set. If you have a set that is no longer serviceable, you might easily be able to remove the 'works' except for the battery-holders and existing loudspeaker and substitute the circuit board of this project for the removed circuits. This would give you a door-alert of neat appearance for very little cost. Existing bell-pushes may be used for S1 and S2, but the illuminated type are not suitable.

9

Light-operated Switch

This switch can be set to operate at any given intensity of illumination. It can be set, for example, to turn on the porch light at dusk. Or it can turn on room lights and a radio set at dusk, even when the house is unoccupied, to give potential intruders the impression that the house is occupied. It can be connected so as to sound a warning on change of light intensity, such as might be caused by furniture or the building catching fire. On a less serious note, you could arrange for it to switch on the radio at sunrise – though the exact time of your being awakened by early-morning music would depend on the season and on the brightness of the morning.

The circuit

The circuit is a simple one (Fig. 9.1). Its action depends on the light-dependent resistor LDR1, sometimes known as a photo-conductive cell. This is a piece of semiconducting material (cadmium sulphide) whose electrical resistance *decreases* as the intensity of light shining on it *increases*. In darkness, its resistance is several hundred kilohms. As the intensity of light increases to that of full sunlight, its resistance gradually drops to only a few hundred ohms. The LDR and the potentiometer VR1 form a potential-dividing chain. The potential at A can be changed either by altering the amount of light falling on LDR1, or by altering the setting of VR1. Given a constant setting for VR1, the potential at A rises if the intensity of light on LDR1 increases. This is the effect of the reduced resistance of LDR1. As the potential at A rises the base current supplied to TR1 increases. In darkness the base current is virtually zero, but above a certain intensity (depending on the setting of VR1) there is sufficient base current to begin to switch TR1 on. As this happens the collector potential of TR1 falls. This reduces the base current to TR2, which is switched off. With no collector current flowing through TR2 and RLA1, the relay coil is not energised; its armature moves to the non-energised position. Conversely, as light intensity falls below a given level, the relay coil is energised. For example,

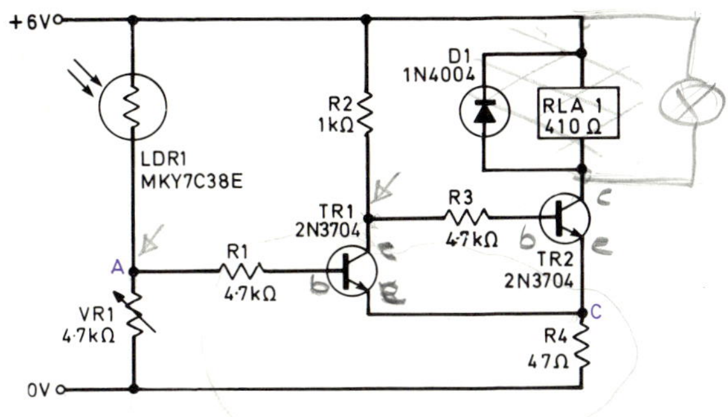

Figure 9.1

Circuit diagram of the light-operated switch

the relay is switched on (energised) at sunset, and can be used to switch on the supply current to a room lamp.

It is essential that slight fluctuations of light intensity around the triggering level do not lead to the relay turning on and off at intervals of a few seconds. For instance, a slight clearing of cloud cover at sunset could lead to a slight but temporary increase of light intensity. If the relay has already been turned on during the earlier stages of sunset, this sudden increase must not be allowed to turn it off again. Once on, the relay should remain on until the following sunrise – or at least until the householder switches off the whole system before retiring to bed. This requirement was found also in the temperature operated alarm (Project 2). The arrangement of TR1 and TR2 in this project is essentially the same as that of TR1 and TR2 in Project 2 (Fig. 2.1), and the way it works is described on p. 6). Diode D1 provides a discharge path for the high voltage induced in the relay coils, as explained on p. 29.

Construction

The circuit can be assembled on a piece of stripboard, as in Fig. 9.2. First solder LDR1, R1, TR1 and R2 in place. Connect VR1 to the terminal pins and temporarily connect a 6V battery to the circuit. A test meter can be used to monitor the potential at point C (Fig. 9.1), which corresponds to the pin at C2 (Fig. 9.2). Shine a bench lamp on LDR1, or arrange it so that it faces the window. As you do this, the voltmeter reading should rise to 4V or more. Shade the LDR with your hand; the reading should fall to 1V or less. With light falling on the LDR, adjust VR1 until the meter reads approximately 1V. Shade the LDR; the meter should now read 0.5V or less. Repeat with the testmeter connected to read the collector potential of TR1; connect the positive terminal of the meter to the wire of R2 that is soldered in hole H15.

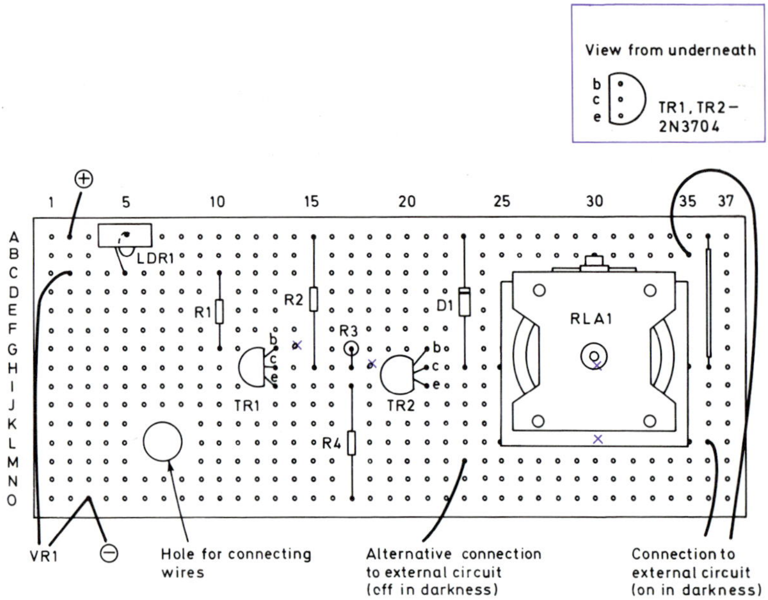

Figure 9.2

Layout of the circuit board for the light-operated switch

With full illumination the reading should be close to zero, but it should rise almost to 6V when the LDR is shaded. Next solder in the remaining components, including the relay. On testing, there should be a click from the relay armature as you shade the LDR with your hand, and another click as you expose the LDR to full light. If this does not occur, check the collector potentials of TR2 these should be high in full illumination and low in shade. If the circuit fails to work at this or any previous stage of testing, check connections, and look for short circuits between copper strips.

A mains power unit for the switch

This circuit can be powered from batteries (6V or 9V) and, if all you require is a circuit to operate a warning lamp (in place of RLA1) or a low-voltage audio alarm (such as the oscillator of Project 2), a battery may be perfectly satisfactory. If you are intending to switch lamps and other devices that are mains powered, it is almost as convenient and certainly more economical to power this circuit from the mains. The circuit for a simple mains power unit is shown in Fig. 9.3. Mains voltage is stepped down to approximately 6V a.c. by the miniature

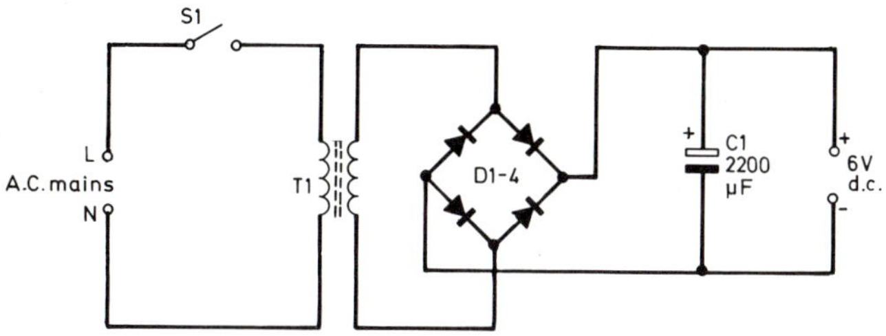

Figure 9.3

Circuit diagram of a simple mains power unit suitable for supplying the light-operated switch and a number of other projects

transformer T1. It is then rectified to 6V d.c. by the bridge (D1–4), and smoothed by C1. This very simple power unit gives a supply that is unregulated and is not precisely 6V, but is suitable for supplying the light-operated switch circuit and also for a number of other projects, as mentioned in other chapters. The components of the power unit may be assembled on a piece of stripboard (Fig. 9.4) the same length as that used for the switch circuit. The two boards may then be slotted into grooves in the wall of the plastics case (Fig. 9.5). Note that the switch circuit is on 0.1in matrix board, but the power unit is on 0.15in matrix board, though it could be mounted on 0.1in board, if you have a spare piece of the correct size.

The two boards are interconnected as shown in Fig. 9.6, in which the live side of mains is shown in red. Since the switch circuit board extends the full depth of the box, a large hole is drilled in it and connecting wires are threaded through this hole before soldering them to the terminal pins. At this stage do *not* connect the switch circuit to the 6V supply. The photograph and drawing show the unit being used to switch power to a table lamp: the lead from the lamp is connected directly to the unit. Alternatively, one or more mains sockets could be bolted to the outside of the case and wired to the relay. Then various lamps or a radio set can be plugged in, as required. If an earth connection is required for the sockets, the mains lead is trimmed to leave its earth wire extra long, so as to extend through the case to the mounted socket.

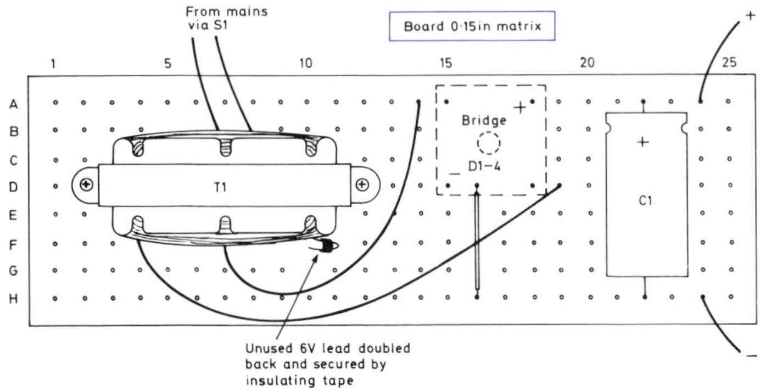

Figure 9.4

Layout of the circuit board for the power unit of Figure 9.3

Figure 9.5

Arrangement of the power unit board (below) and switch circuit board (above) in the case. The LDR can be seen at the edge of the upper board, on the right; the piece of translucent plastics sheet that covers the opening in the panel can be seen on the upper left side of the panel

The potentiometer is mounted on the front panel. A hole about 1cm in diameter is punched in the panel to allow light to reach the LDR. As shown in Figs. 9.2 and 9.5, the LDR is mounted so that its receptive surface faces toward the front panel when the board is slotted in

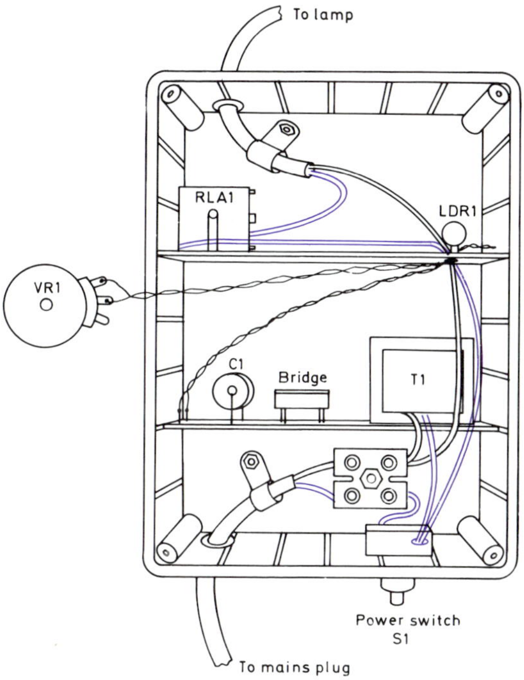

Figure 9.6

Connections to the mains supply and between circuit boards

position. To protect the LDR a rectangle of transparent or translucent plastics sheet may be fixed behind the hole, using contact adhesive. In the prototype a piece of yellow PVC about 2mm thick was cut from an old slide box.

The unit is first tested with the boards slotted into the case but before fixing the panel in position. Switch on the power switch. Check that the output from the mains unit is approximately 6V. While checking, remember that exposed terminals, terminal screws and most of the exposed parts of the relay are *live*. If you touch these, you could receive a serious electric shock. If the power output is excessively high (more than 10V), or is very low, switch off and remove the mains plug from its socket. Check the mounting of the components on the power unit board; check the connections to the board. The only component likely to be at fault is the bridge; use a test meter to check this. It is also possible that one of the transformer windings has burnt out, especially if you are using a transformer taken from other older equipment. This fault would result in no power output from the unit.

Figure 9.7

The completed light-operated switch

When the power unit is working properly, switch off and remove the mains plug from its socket. Then solder the connecting wires from the switch circuit board to the power unit output pins. You have already tested the switch circuit so, if the mains connections to the relay are properly made, the whole unit is now ready for action. Fix the front panel in position; as you do so, look to see that the potentiometer does not make electrical contact with any other component (especially the relay). Check also that the hole in the panel is correctly aligned with the LDR. The completed unit is shown in Fig. 9.7.

Using the switch

Insert a lamp in the lamp socket connected to the unit. The lamp should be at some distance from the unit, or shaded so that its light does not fall directly on the unit. Shine another lamp on the unit, or place the unit beside a window, through which daylight can fall on it. Turn the knob to its fullest extent anticlockwise (to LO, Fig. 9.7). The controlled lamp should now be *on*. Turn the knob slowly clockwise (toward HI) until the lamp goes *off.* While doing this, take care that your hand does not shade the LDR. Now place your hand over the LDR hole; the lamp should turn *on* again. Remove your hand; it should go off.

You can now experiment to see what happens if the light from the controlled lamp is allowed to fall directly on the LDR. As you might expect, the light from the lamp is detected by the LDR and this causes

Table 9.1. Components list for the light-operated switch

Resistors	
R1, R3	4.7kΩ (two off)
R2	1kΩ
R4	47Ω
all carbon, $^1/_4$W	
Potentiometer	
VR1	4.7kΩ, linear
Semiconductor	
TR1, TR2	2N3704
D1	1N4004
LDR1	MKY7C38E or ORP12 light dependent resistor (cadmium sulphide cell)

Miscellaneous

0.1in matrix stripboard 15 strips by 37 holes; 0.1in matrix terminal pins (six off); RLA1, miniature open printed-circuit relay, 6V, 410Ω coil; knob for potentiometer; plastics case (for example a.b.s. case, type M3; connecting wire; solder: scrap transparent or translucent plastics sheet: contact adhesive

Table 9.2. Components list for mains power unit and associated mains connections

Capacitor	
C1	2200μF 16V electrolytic
Semiconductors	
D1–4	WO4 bridge rectifier (1 off)

Miscellaneous

0.15in matrix stripboard 8 strips by 25 holes; 0.15in matrix terminal pins (two off); T1 subminiature mains transformer, mains voltage primary, 6-0-6V secondary 100mA (Brazenose type 606/1, Electrovalue Ltd for UK mains voltage); case as above; S1 Castelco switch type 2634, single pole push off push on; dual terminal block; cable grips (two off); grommets (two off); 6 B.A. bolts, nuts and shakeproof washers (three off each); mains leads, mains plug; connecting wire; solder

the lamp to be switched off. Immediately this happens, the resulting lack of light is detected by the LDR, causing the lamp to be switched on again. Then the whole cycle is repeated, several times per second. The lamp flickers in a useless fashion. This effect must be remembered when setting up the unit to switch on a room lamp at sunset. The unit is best placed on a window-sill with the LDR facing toward the window. The lamp is placed in the room at some distance from the unit. Preferably a curtain is drawn across to shade the unit from the lamp.

10

Christmas Star

This is a novel and fascinating decoration for your Christmas Tree. It consists of a star-shaped base decorated with tinsel and 'glitter' (Fig. 10.1) which has a small red lamp at each point of the star. The red lamps are light-emitting diodes that are made to turn on and off in a definite

Figure 10.1

The Christmas star sparkles in a fascinating way among the other decorations on the Christmas tree

sequence by the circuit hidden behind the star. Although the sequence is actually a regular one it appears on casual inspection that the lamps are flashing in a slow, random manner.

How the multivibrator works

The flashing of the lamps is produced by an astable multivibrator (Fig. 10.2). In Project 3 we built an astable multivibrator by using logic gates, and used it to flash the lamp of the Flashing Messenger.

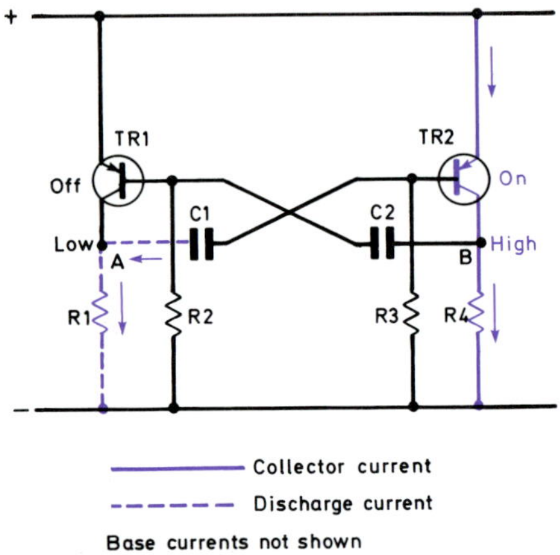

Figure 10.2

The action of an astable multivibrator circuit

Here we are building the multivibrator from a pair of transistors, a pair of capacitors and a few resistors. The circuit used here has some points in common with the bistable circuit of Fig. 6.1 and, like that circuit, can exist in two states. Unlike the bistable, which remains permanently in one state until it is triggered to change, the astable changes state automatically after a fixed period of time. One state of the astable multivibrator is shown in Fig. 10.2, where TR1 is off and TR2 is on. This means that the potential at point A is low and that at point B is high. When the astable entered this state, the potential at A suddenly changed from high to low. This negative pulse was transmitted through C1 to the base of TR2, turning TR2 on. Immediately TR2 was turned on, the potential at B went high, and a positive pulse was transmitted through C2 to the base of TR1, turning it off. This is the state shown in the figure. In this state C2 begins to discharge through R2. Gradually the base potential of TR1 falls, the rate of fall depending on the values of C2 and R2. Eventually the potential at the base of TR1 becomes low enough to draw a base current sufficient to bring TR1 into a conducting state. This causes a rise in potential at A, which is transmitted through C1 to the base of TR2. This is partly turned off, causing a fall of potential at B. This fall is transmitted through C2 to the base of TR1, which is then turned even more fully on, causing the potential at A to fall very low, switching TR2 fully off. In this way the astable has changed to its other state, with TR1 on, TR2 off, A high, and B low. Then C2 begins to discharge through R3 and, after a given period of time, the circuit reverts suddenly to its original state.

The circuit

In the Christmas Star circuit (Fig. 10.3) C1 and C2 have high capacitance, giving a relatively low rate of change of state. As the multivibrator oscillates, the potential of the collector of TR2 is alternately high and

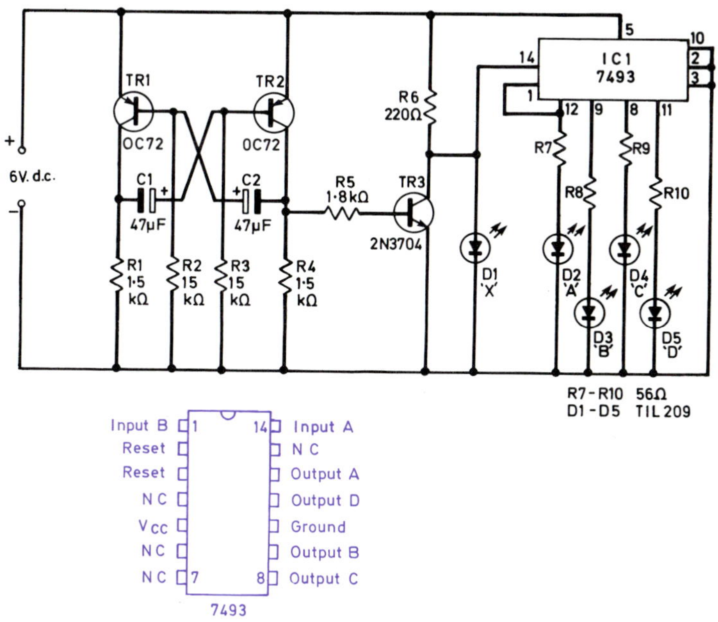

Figure 10.3

Circuit diagram of the Christmas star

low. When it is low, TR3 is off and the potential at its collector is high, causing a current to flow through the light-emitting diode D1, otherwise known as lamp X. The result is that lamp X flashes on and off about once a second.

The potential at the collector of TR3 also becomes the input for the integrated circuit IC1. This is a 4-stage binary counter, which performs its function in the following manner. The alternating potential from TR3 is fed to input A (pin 14). Every time input A goes low, output A changes state. If output A was low it goes high, if it was high it goes low. Output A (pin 12) is connected through R7 to D2, another LED, to be referred to as lamp A. When output A goes high, lamp A lights otherwise it is dark. We can now predict what happens as lamp X flashes on and off. The sequence of states of the two lamps is as follows:

Step no	*Lamp A*	*Lamp X*
0 (start)	0	0
1	0	1
2	1	0
3	1	1
4	0	0
5	0	1
6	1	0
7	1	1
8	0	0 and so on. . .

In the table a '0' represents a dark lamp (corresponding to low), and a '1' represents a shining lamp (corresponding to high). The sequence is repeated every 4 steps. If you are familiar with binary arithmetic, you will know that the sequence of states of the lamps in the table above corresponds to the numbers 0, 1, 2, 3 written in binary notation – 00, 01, 10, 11. If you run your eyes down the columns you will also see that lamp A flashes at half the rate of lamp X.

Output A is also connected to input B (pin 1), and stage B behaves in a similar manner to stage A. Its output changes state whenever its input goes low. There are also internal connections to two further stages in the counter, stages C and D. These too behave in the same manner as stages A and B. The effect of this is that outputs and the lamps connected to them go through a 32-step sequence corresponding to the numbers 0 to 31 in binary notation:

Step no	*Lamps*					*Number of lamps lit*
	D	*C*	*B*	*A*	*X*	
0	0	0	0	0	0	0
1	0	0	0	0	1	1
2	0	0	0	1	0	1
3	0	0	0	1	1	2
4	0	0	1	0	0	1
5	0	0	1	0	1	2
6	0	0	1	1	0	2
7	0	0	1	1	1	3
8	0	1	0	0	0	1
9	0	1	0	0	1	2
10	0	1	0	1	0	2

Step no	*Lamps* *D*	*C*	*B*	*A*	*X*	*Number of lamps lit*
11	0	1	0	1	1	3
12	0	1	1	0	0	2
13	0	1	1	0	1	3
14	0	1	1	1	0	3
15	0	1	1	1	1	4
16	1	0	0	0	0	1
17	1	0	0	0	1	2
18	1	0	0	1	0	2
19	1	0	0	1	1	3
20	1	0	1	0	0	2
21	1	0	1	0	1	3
22	1	0	1	1	0	3
23	1	0	1	1	1	4
24	1	1	0	0	0	2
25	1	1	0	0	1	3
26	1	1	0	1	0	3
27	1	1	0	1	1	4
28	1	1	1	0	0	3
29	1	1	1	0	1	4
30	1	1	1	1	0	4
31	1	1	1	1	1	5
0	0	0	0	0	0	0 and so on

The lamps flash at different rates – each at half the rate of the lamp before it in the sequence. Gradually but erratically the number of lamps that are lit increases, until at step 31 they are all alight. Then they all go out together, and the cycle begins again.

Construction

The circuit is assembled on a small piece of stripboard Fig. 10.4. The layout is compact so that the board may be easily hidden at the back of the star. First build the multivibrator (TR1, TR2 and the resistors and capacitors connected to them). This may then be tested by connecting a voltmeter between the negative rail and the collector of either of the transistors. Voltage should show a regular change from low (almost zero) to high (almost 6V) about once a second. It should remain in its high and low states for approximately equal lengths of time. Then connect TR3, R6 and temporarily connect D1. When tested, D1 should flash at a rate of approximately 1Hz. If you prefer, you may use a socket for IC1, but as only 10 of its pins need to be soldered to the board, it is not too difficult to remove this i.c. should a fault be discovered later. Solder the pins one at a time, with an interval of a minute or two between solderings, so as not to overheat the i.c. Note that the i.c. is mounted 'upside-down', with its index end (pins 1/14) toward the lower edge of the circuit board. This is the opposite orientation to the conventional one, as used in Fig. 10.3. Note also that pins 2 and 3 are connected together. They are grounded through the connecting wire from hole E24 to C19. These pins are the inputs to the 'reset' circuit. The counter operates only when both of these inputs are low; if either is high, the counter resets and all outputs go low and remain low.

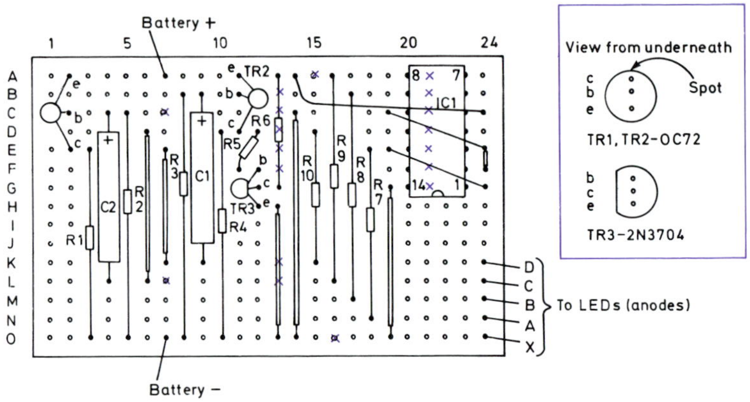

Figure 10.4

Layout of the circuit board for the Christmas star

The circuit may be finally tested by applying a voltmeter (or more than one if you have several) to the output pins at K24, L24, M24, N24 and O24 and checking that the correct sequence of highs and lows is obtained, as set out in the table above.

The star is cut from hardboard. Draw a circle on the board, about 10cm in *radius*. At intervals of exactly 72° around the circumference of the circle mark the points of the star. Join these by straight lines to obtain a shape like that of Fig. 10.5. Bore small holes near the points, to take the LEDs. The holes should be just a little less in diameter than the diameter of the rim of the LED, so that the LED rests on the front surface of the board (see sectional view in Fig. 10.6). The leads pass through the hole and protrude from the back of the board. With the TIL209 the anode lead is identified as the shorter lead. Fix the LEDs in place by placing a drop of quick-setting resin (Araldite) in the hole. Thickly coat the centre of the rear surface with resin and press the circuit-board firmly into this (Fig. 10.7). Leave the assembly to harden – overnight, if necessary. The cathode of each LED is connected to the negative terminal pin (O7). The anodes are each connected to the pins at K24–O24.

Figure 10.5

Front view of the Christmas star

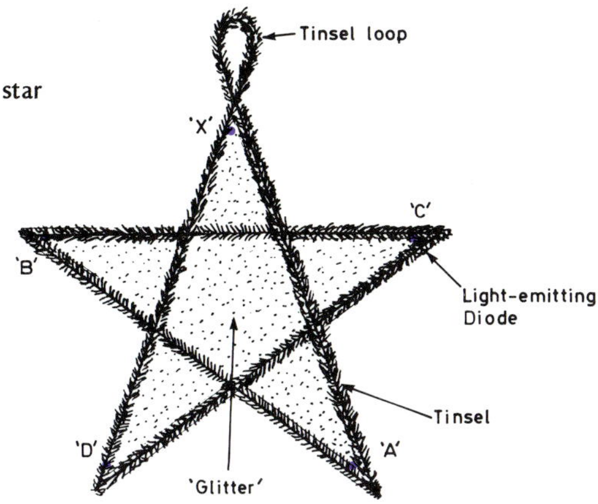

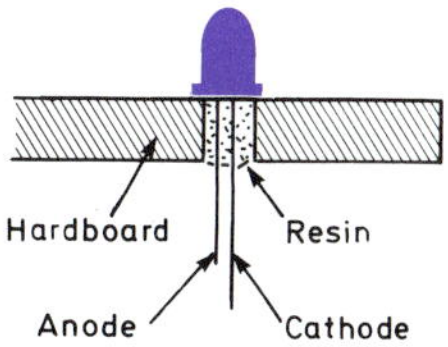

Figure 10.6

Mounting the light-emitting diodes on the star

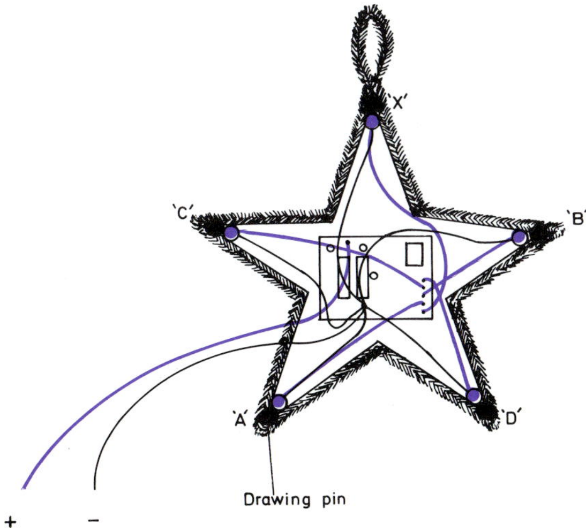

Figure 10.7

Rear view of the star showing the connections to the battery and to the LEDs

Note the layout of the LEDs (Fig. 10.5); this arrangement gives a pseudo-random effect. The star is decorated with tinsel and 'glitter'. Drawing pins are first pressed into the rear surface of the hardboard, as close as possible to the points of the star. A continuous length of tinsel is then wound around one pin, then across the front of the star to the next point but one. Here a turn around the pin holds the tinsel in place, and then on to the next point but one, and so on, finishing with a loop at point 'X' with which to hang the star from the tree.

Table 10.1. Components list for Christmas Star

Resistors	
R1, R4	1.5kΩ (two off)
R2, R3	15kΩ (two off)
R5	1.8kΩ
R6	220Ω
R7–R10	56Ω (four off)
all carbon, $^1/_4$W	
Capacitors	
C1, C2	47μF, 10V electrolytic
Semiconductors	
TR1, TR2	OC72 (two off)
TR3	2N3704
D1–D5	TIL209 (red, or other colours if preferred) (five off)
Integrated circuit	
IC1	7493 TTL 4-stage binary counter
Miscellaneous	
0.1in matrix stripboard, 15 strips by 24 holes; 0.1in matrix terminal pins (seven off); hardboard; drawing-pins (five off); length of tinsel (approx 1 metre); tube of 'glitter'; epoxy resin adhesive; crocodile clips or other connections to battery; battery holder for four SP11 cells; Christmas wrapping paper; connecting wire; solder	

Areas of the front of the star that show between the tinsel strands are coated in adhesive and 'glitter' is thickly sprinkled on. When the glue is dry, the surplus glitter may be tipped off. Long leads are then run from the pins at A7 and O7 to form battery connections. The battery may be four SP11 cells in a plastics battery holder. Connection to this holder is made by crocodile-clips or press-studs. Alternatively, use a 6V battery of the type used with electric bell systems. Batteries of the types suggested will provide power for several hours – normally sufficient to last the whole Christmas season. The battery may be wrapped in Christmas wrapping-paper, leaving a small gap to give access to the terminals. When so disguised, it resembles a Christmas parcel, that may be hung from a stout branch at a short distance from the star.

11

Process Timer

The timer can be set to run for any whole number of minutes, from one to five, and a flashing lamp indicates when the set time has elapsed. Not only can it be used for timing boiling eggs, but also for timing other periods of short duration, such as those involved in photographic processing, and competition scrabble or chess.

The timer circuit

The circuit is based on a single i.c., the 556 dual timer i.c. This contains two separate timing devices, each identical to the 555 timer used in

Figure 11.1

The circuit in use as an egg timer

Project 7. The design is a compact one so that the completed timer is small enough to perch on top of the cooker (Fig. 11.1) or to occupy a small space next to your plate at breakfast, where it will not get in your way, yet gives instant warning that the eggs have finished cooking.

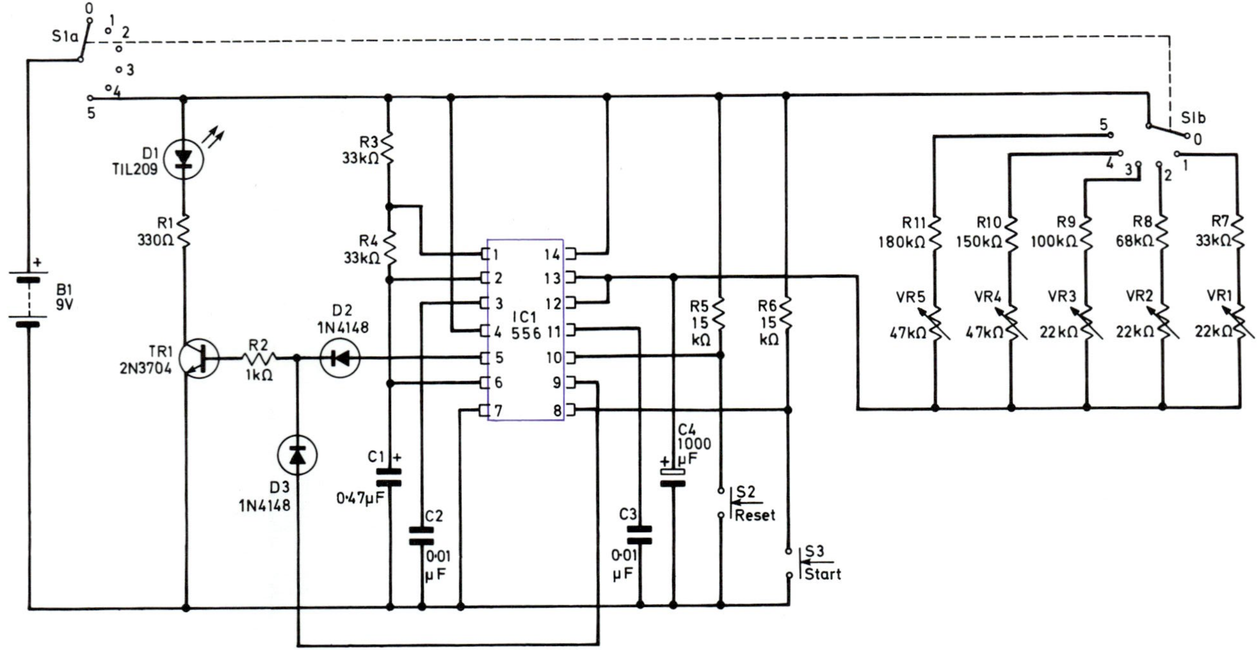

Figure 11.2

Circuit diagram of the timer

One of the timers in the i.c. is connected as a monostable multivibrator, (described in Project 7) to give a high output for a period of 1, 2, 3, 4 or 5 minutes after the start button is pressed. In Fig. 11.2, C4 is the timing capacitor for this circuit. The timing resistors consist of five fixed resistors, each wired in series with a variable resistor. By turning the rotary switch S1, any one of the fixed resistors and its corresponding variable resistor is selected and connected to the timing circuit. The values of the resistors are chosen to provide periods of the correct length, with the variable resistors being used to adjust the periods exactly.

The other timer is connected as an astable multivibrator. Its timing capacitor is C1. It is charged by the current passing to it through R3 and R4. During this period, output (pin 5) is high. When C1 has been charged to + 6V (i.e. $\frac{2}{3}$ of 9V), this state is detected by the threshold input (pin 2), and the capacitor is then discharged (through R4 only), to pin 1. At the instant that discharge begins, output (pin 5) falls to zero. The capacitor is eventually discharged to about +3V (i.e. $\frac{1}{3}$ of 9V) at which level the trigger input (pin 6) is activated to trigger the circuit; output goes high again and the cycle is repeated indefinitely. The reset input (pin 4) is not used in this mode of operation and is permanently connected to the 9V rail. The rate of repetition of the cycle is governed by the rates of charge and discharge of C1, which are dependent on the values of C1, R3 and R4. With the values chosen, the timer oscillates at about 2Hz.

The outputs of both timers are fed via diodes D2 and D3 and resistor R2 to the base of TR1. The output of either timer alone is sufficient to turn TR1 fully on. Before timing is begun, the output from pin 9 is low, and the output from pin 5 is alternating between high and low at 2Hz. Consequently, TR1 is turned on and off at the same rate, and the light-emitting diode D1 flashes. As soon as the start button is pressed, the output at pin 9 goes high. Now a continuous base current flows to TR1, and D1 shines continuously. As soon as the timed period is over, the output from pin 9 goes low again, and the lamp begins to flash. Thus the lamp is on during the timing period and flashes before and after the timing period. The flashing light warns you when the timing period is complete, and also reminds you to switch off the timer to conserve the battery.

Construction

The circuit is assembled on a single piece of stripboard (Fig. 11.3), which fits upside-down into a small plastics case (Fig. 11.4). First solder the i.c. in position (use a socket if you wish), and then the components needed for the astable multivibrator. These are R3, R4, C1 and C2. Both

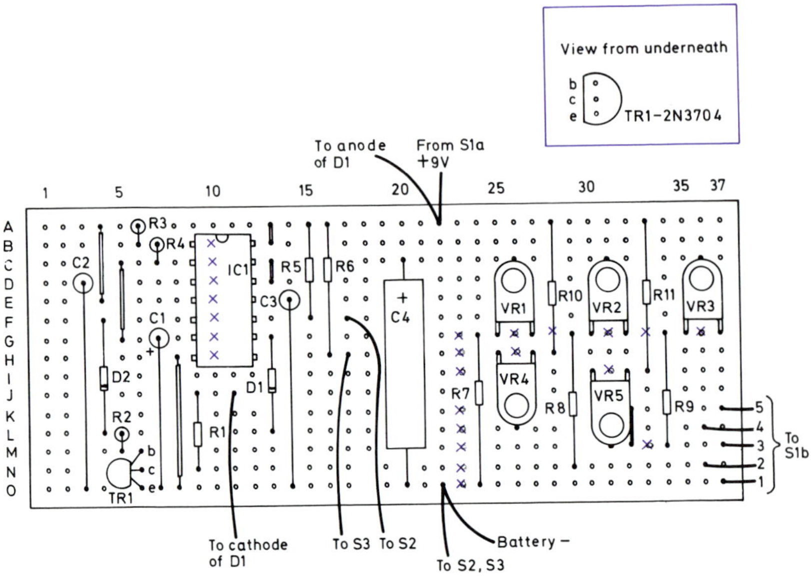

Figure 11.3

Layout of the circuit board for the egg timer

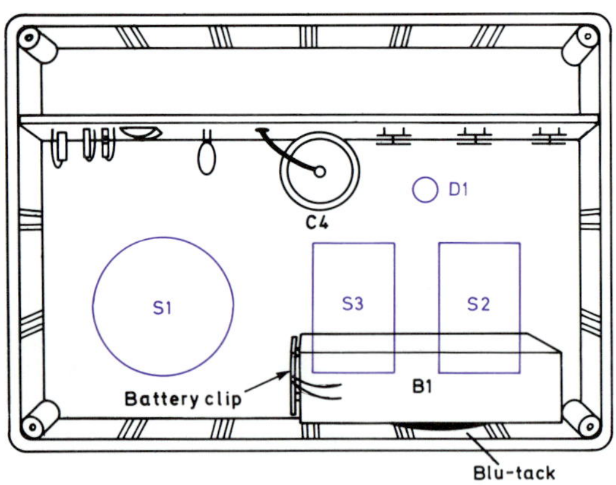

Figure 11.4

The arrangement of the circuit board and components in the case. The items shown in red are panel-mounted, and fit in the space below the board as shown. The battery is fixed toward the right to allow room for S1 on the left

circuits share power connections to pins 7 and 14. The connection to pin 7 is by wire, but that to pin 14 is by a blob of solder joining strip A to strip B near their 13th holes. Make a wire connection between the reset input (pin 4) and the positive rail (A4 to E4). For testing, connect a battery to the terminal pins at A22 and O22. Use a voltmeter to measure the output from pin 5. The needle should flicker rapidly and regularly around the 4.5V mark.

Next solder D2, R2, TR1 and R1 in place. Temporarily wire D1 between the terminal pin at J11 (cathode) and the positive rail (anode). When the battery is connected, D1 should flash at about 2Hz.

Now assemble the monostable multivibrator. Solder C3, C4, R5, R6 and D3 in position. Temporarily solder a resistor of value about 10kΩ between A24 and C24. This enables you to check the operation of the monostable *before* soldering the fixed resistors R7–R11 and potentiometers VR1–VR5 in position. With a 10kΩ timing resistor, the monostable should go high for about 5 seconds. Use a short length of wire to connect the terminal pin at H17 momentarily to the OV rail. This should set the timer; its output should go high. D1, which was

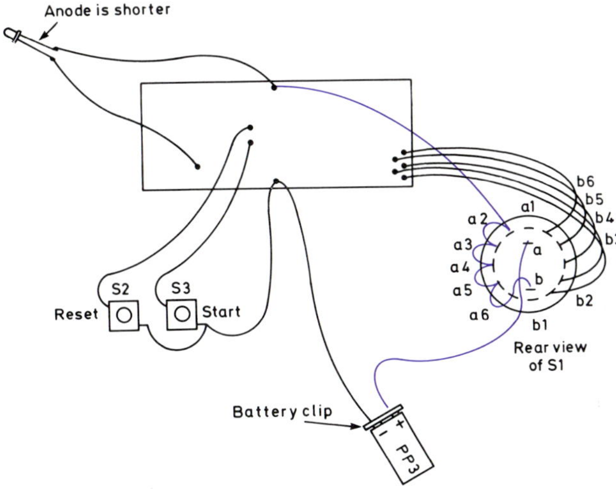

Figure 11.5

Connections between the circuit board and other components

previously flashing, should now shine continuously (though a *slight* pulsation in its intensity is detectable with careful observation). After approximately 5 seconds, D1 should begin to flash again. Start the timer once more, but before 5 seconds have elapsed, momentarily short the reset terminal pin (at F17) to the OV rail. The lamp should immediately begin to flash.

Assuming that all is in working order, remove the temporary resistor. Solder the five timing resistors and their potentiometers in place, taking special care that the copper strips are cut at exactly the correct places. It is best to continue with the assembly of the panel-mounted components (Fig. 11.5). Keep the leads between board and panel short (for there is not much spare space inside the case) yet leave them long enough to allow you to manoeuvre the board and battery into their positions within the case.

The times are adjusted before putting the device into its case. Begin with S1 in position 1; this brings R7 and VR1 into circuit. Turn VR1 as far anticlockwise as it will go. Press S3 and, using a watch or clock with a second hand, measure the length of time for which D1 shines steadily. This tells you the minimum time possible with S1 in position 1. Next turn VR1 as far clockwise as it will go; press S3 and thus find the maximum time for position 1. It is to be expected that the minimum time will be less than one minute and the maximum time more than one minute. If this is so, it is easy to estimate how far VR1 must be turned to obtain a time of *exactly* one minute. Turn VR1 to the estimated position, press S3 and time again. If any further adjustments are necessary, these can be made and the timing checked until you are satisfied that the timer is set accurately enough for the purpose. If you are cooking eggs, an error of 5 seconds is acceptable!

If by chance the minimum period comes to *more than* 1 minute, R7 must be removed and replaced by a resistor of lower value. Similarly, if the maximum period comes to *less than* 1 minute, R7 must be removed and replaced by a resistor of higher value. The reason for the required time being outside the settable range is probably that C4 has a value widely different from 1000μF. Electrolytic capacitors have a tolerance of – 20% to + 50% and, on occasion, we may find one that departs widely from its stated value.

The instructions above should now be repeated with the switch in position 2, adjusting VR2 to obtain a time of exactly 2 minutes (to within 5 seconds). In the same way adjust VR3, VR4 and VR5 to obtain times of 3, 4 and 5 minutes respectively. An error of 10 seconds may be tolerated in the longer times. Adjustments for the longer times are rather tedious and time-consuming, but fortunately need be done only once.

When adjustments are complete, slide the board into the groove in the walls of the case. Fix the battery in position, using a small lump of Blu-Tack, or one of those foam pads, self-adhesive on both sides, that are sold for fixing pictures to walls and all manner of other items to all manner of surfaces. Gently place the panel in position, making sure that none of the components on it makes electrical contact with the circuit board. Fix the panel in place with its retaining screws.

Current consumption is low and the unit is used for only a few minutes each day, so a PP3 battery should last for several months. In use the switch is turned from the *off* position to one of the numbered positions marked on the panel (Fig. 11.1); the lamp begins to flash. When you are ready to begin cooking the egg, check that the switch is turned to the number of minutes of cooking time, begin cooking and

Table 11.1. Components list for the timer

Resistors	
R1	330Ω
R2	1kΩ
R3, R4	33kΩ (2 off)
R5, R6	15kΩ (2 off)
R7	33kΩ
R8	68kΩ
R9	100kΩ
R10	150kΩ
R11	180kΩ
all carbon $^1/_4$W	
Potentiometer	
VR1, VR2, VR3	22kΩ
VR4, VR5	47kΩ
all miniature horizontal preset potentiometers (type PR10, Electrovalue Ltd)	
Capacitors	
C1	0.47μF tantalum
C2, C3	0.01μF ceramic disc (two off)
C4	1000μF electrolytic, 10V
Semiconductors	
TR1	2N3704
D1	TIL209, light-emitting diode, red
D2	1N148 (two off)
Integrated circuit	
IC1	NE556, dual timer
Miscellaneous	
0.1in matrix stripboard, 15 strips by 37 holes; terminal pins for 0.1in stripboard (ten off); S1, 2-pole, 6-way, rotary switch; knob for S1; S2, S3 push-button switch, single pole push to make, Castelco type 2644 (Electrovalue Ltd, two off); battery press-stud connector twin type for PP3 battery; a.b.s. plastics case type M2	

press the start button. The lamp then shines steadily and remains on until the correct time has elapsed. If by any chance you make a mistake and want to begin again, press the reset button. The lamp begins to flash and timing can then be started again. As long as the lamp remains continuously on, timing is in progress. When it begins to flash, the period is complete.

12

Power Pack

Many items of electrical equipment, such as radio sets, tape recorders, record players, calculators, and even TV sets, are portable nowadays, thanks to the invention of the transistor. Many of these items are small enough to fit inside the pocket. True portability brings several advantages but has one main snag. Portable equipment is powered by batteries, and batteries are very expensive. This is particularly felt by those who run equipment with high power consumption, such as record players and tape recorders. The solution to the problem of cost is to use batteries only when you must – when using the equipment in the car or on a picnic or a trip to the beach. When at home, money can be saved by removing the batteries and obtaining power from a mains power pack instead. Several circuits for obtaining rectified d.c. from mains current have been described in this book in conjunction with other projects. These are relatively simple circuits and have their limitations: the voltage supplied cannot be set exactly; voltage drops occur if a heavy load is connected to the output; the output voltage changes if mains voltage changes; the output has a ripple that will produce a hum if used with audio equipment. Another disadvantage of the simple type of converter is that should a short-circuit develop in the equipment being powered, high currents can be drawn from the converter and these may cause extensive damage to the equipment. The power pack described in this project, as well as giving an accurately determined and stable voltage at its output, has built-in current limitation; if the resistance of the load is low (through a short-circuit, for example) and the current flowing to the equipment becomes excessively high, the power pack automatically reduces its output voltage to a very low level.

Circuits that provide a regulated, stabilised, current-limited supply can be fairly complex to build. Here we simplify construction by making use of an i.c. specially designed to perform these functions. This is the 723 variable voltage regulator i.c. This versatile i.c. can be used in a number of configurations. First we show how it is used as the basis of a power pack giving 9V d.c. output. Later we describe modifications of the design to suit other applications.

The circuit

The complete circuit for the converter is given in Fig. 12.1. The mains supply is transformed to approximately 18V a.c. by T1, then rectified to 18V d.c. by the bridge (D1–4). Capacitor C1 has high capacitance,

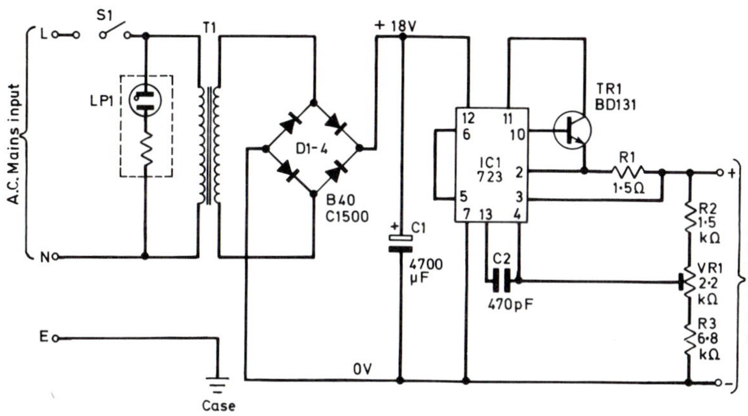

Figure 12.1

Circuit diagram for a power pack supplying 9V d.c. with current limiting

smoothing the rectified current adequately. Power is supplied to the i.c. through pins 11 and 12; pin 7 is the connection to the negative rail. The regulated output is obtained from pin 10. The i.c. is capable of supplying up to 150mA, but this is not sufficient for driving equipment such as record players and tape recorders. These require currents of the order of 300 or 400mA. We therefore use the output from pin 10 to supply base current to a power transistor, TR1. This relatively small base current controls the flow of the relatively large collector current that is drawn directly from the positive rail, passes through TR1 and R1, and continues to the positive output terminal. We will discuss the function of R1 later. Between the positive and negative output lines there is a chain of resistors of high value. The current passing along this chain is less than 1mA. The chain acts as a potential divider and a means by which the i.c. is informed of variations in its output voltage.

The i.c. gives a *stable* output when the potential applied to pin 14 is exactly equal to the potential applied to pin 5. Pin 5 is connected to pin 6, which is internally connected to a zener diode, giving a reference voltage of approximately 7V. Thus pin 5 is at 7V and, for a stable output, pin 14 must be held at 7V too. This is achieved by adjusting the setting of VR1 so that when the potential across the whole chain of resistors is exactly the required output voltage, 9V, the potential at the wiper of VR1 is exactly 7V. If the potential at the wiper is exactly 7V, the potentials at pins 4 and 5 are equal. If they are equal, the output potential at pin 10 remains at a steady value – it is stabilised. At what value is it stabilised? It is stabilised at a value that will produce sufficient base current to turn TR1 on by just the right amount to maintain a potential of exactly 9V across the resistor chain.

Suppose that we connect a low-resistance load across the output terminals – a tape recorder, for example, that takes a current of 350mA. The effect of this would be to cause a drop of potential across the resistor chain. The potential at the upper end of R2 would fall below 9V. Consequently, the potential at the wiper of VR1 would fall below 7V. An amplifying circuit in the i.c. detects the fall of potential at pin 14 relative to the steady reference voltage applied to pin 5. The difference is amplified internally and results in an increase in the output potential at pin 10. This produces a greater base current, TR1 is turned more fully on and more current flows to the load (the tape recorder). The amount of current supplied is increased until it is large enough to supply the tape recorder yet still produce a potential difference of 9V across the resistor chain. When this state is reached, the wiper of VR1 is once more at 7V, the inputs to pins 14 and 5 are once again equal, and the output at pin 10 stabilises at this new value. In this circuit we have a good example of negative feedback. The i.c. detects a falling potential at pin 14 and responds by producing a rising potential at pin 10. We use this effect to obtain a *stabilised* output voltage.

To find the output voltage V_o, call the reference voltage V_{ref} (which in the 723 is 7V) and let R_a be the resistance between the positive end of the chain and the wiper of VR1, and let R_b be the resistance between the negative end of the chain and the wiper of VR1:

then $$\frac{V_o}{V_{ref}} = \frac{R_a + R_b}{R_b}$$

This equation depends on the fact that since the same current flows through all resistors in the chain (ignoring the minutely small current flowing to pin 14) the potential drops across the whole chain or parts of the chain are in proportion to the resistance of the whole or parts.

For our circuit, $\dfrac{9}{7} = \dfrac{10\,500}{R_b}$

From which we calculate $R_b = 10\,500 \times \dfrac{7}{9} = 8167\Omega$

This means that the total resistance of R3 and the portion of VR1 between its wiper and its negatively connected end must be 8167Ω. Since R3 is 6800Ω, the portion of VR1 between the wiper and its negative has resistance 1367Ω – the setting is just a little more than half-way along the scale of VR1.

Suppose we now turn VR1 so that its wiper is exactly at its negatively connected end. This makes $R_b = 6800\Omega$, and $(R_a + R_b) = 10\,500\Omega$, as before. The i.c. automatically adjusts its output to bring the *wiper* potential to 7V, as before, since this potential must be exactly equal to V_{ref}, which is fixed at 7V. Now we can calculate what the new output voltage will be:

In the equation $\dfrac{V_o}{7} = \dfrac{10\,500}{6\,800}$

From which we calculate $V_o = \dfrac{10\,500}{6\,800} \times 7 = 10.8\text{V}$

The output voltage is now stabilised at the new value of 10.8V. Similarly, if we turn the wiper of VR1 as far as it will go toward its positively connected end, we can calculate the expected output voltage (the voltage needed to produce 7V at the wiper):

$$V_o = \frac{10\,500}{9\,000} \times 7 = 8.2\text{V}$$

These calculations show us that the circuit can deliver a stabilised output voltage of any value between 8.2V and 10.8V.

We have now seen that the output is *regulated* (can be set to some definite value within a given range) and is *stabilised* (holds its set value independently of variations in the mains supply voltage and variations in the load placed across the output terminals). The other feature to consider is the ability to *limit its output current.* pins 2 and 3 are connected to either end of the low-value resistor R1. As the output current increases, the potential drop across this resistor increases. When the potential difference between pins 2 and 3 exceeds a certain value, V_{max}, the output potential at pin 10 is automatically lowered – the output current is thus reduced to a low level. The shut-down begins when the potential difference exceeds about 0.7V. Thus if R1 is 1.5Ω, we can use Ohm's law to calculate that the maximum current I_{max} is:

$$I_{max} = \frac{V_{max}}{R} = \frac{0.7}{1.5} = 0.47\text{A (470mA)}$$

The output current is limited to about 470mA, which is sufficient to drive most portable record players and tape recorders. If you do not intend to use the unit with such equipment, you can limit the current to some lower value, and so give greater protection to your equipment in the event of its failure. For example, if R1 has the value 3.9Ω, the maximum current is

$$I_{max} = \frac{0.7}{3.9} = 0.18\text{A}$$

This is adequate for most portable radios.

Construction

The prototype was built into a tea tin. The shape is convenient for enclosing a transformer, with a little space to spare for the remainder of the circuit. However, if you prefer to use a more conventional case, you will gain the advantage of having more room in which to manipulate the soldering iron, but the finished unit will lack the convenient compactness of the prototype (Fig. 12.2). The transformer is to be bolted to the bottom of the tin; a solder tag on one bolt is to be soldered to the earth lead of the mains cable (Fig. 12.3). First drill holes in the tin for the transformer bolts and the mains cable and grommet. The switch S1 and indicator lamp LP1 are mounted in holes bored in the side of the box. About 10 holes about 4mm in diameter are bored near the top of the opposite side of the tin, and a further 10 holes at the bottom of that side, to allow ventilation. At this stage the tin may be painted. A pleasant and serviceable finish may be obtained by spraying it with crinkle-finish paint. Spray the lid too – holes will be needed in it, but these can be drilled later.

Next mount the transformer, switch and lamp and make the necessary connections between them. Connect the mains cable – not forgetting the earth connection.

The three secondary leads of T1 correspond to 9, 0 and 9V respectively. The central lead (0) is not required; cut it fairly short and bind the end with insulating tape. The other two leads are, for the moment, taken out through two of the holes at the top of the tin. Place the lid in

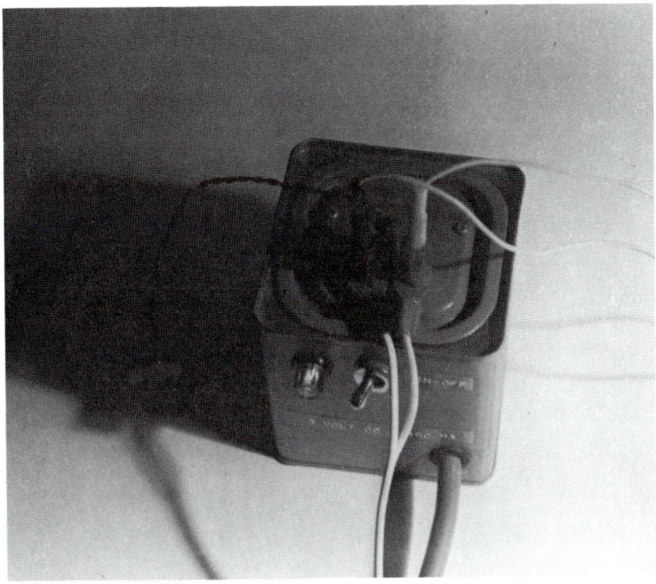

Figure 12.2

The completed power pack, showing various types of connection – *on the left* 2-pin socket and battery clip (see text); *centre* spade terminals; *on the right* 4 mm plugs

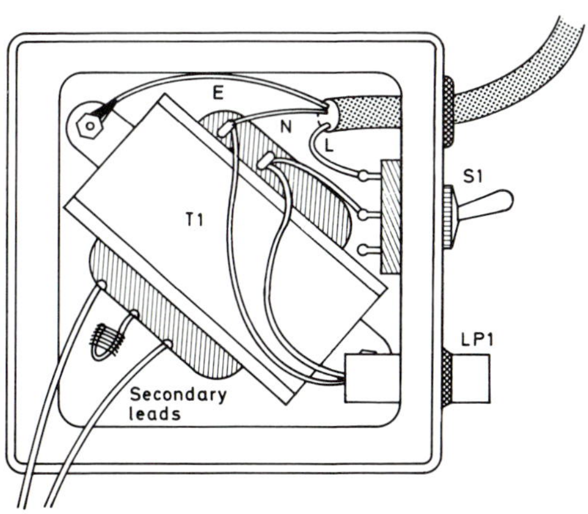

Figure 12.3

The arrangement of the transformer in the case, and the connections to it. Note the central secondary lead, cut short and bound with insulating tape

position. You now have a temporary and reasonably safe 18V a.c. supply to use when testing the regulator circuit.

The regulator circuit is assembled on a piece of stripboard, as in Fig. 12.4. Solder the bridge in place first, and check that it rectifies correctly. When connected to the 18V a.c. supply, the output from the bridge should be 18V d.c. Next solder the remainder of the components in position, taking care to make no errors. Remember the solder connections beneath the board, joining pin 5 to pin 6, and pin 11 to pin 12. When all components are in position and checking is complete, connect the board to the 18V a.c. supply. Connect a voltmeter between

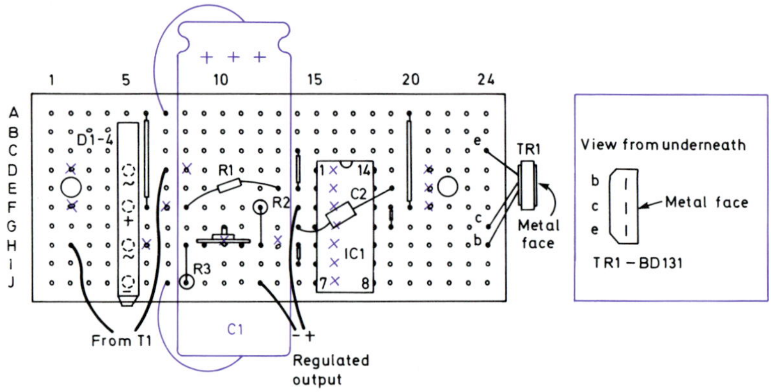

Figure 12.4

Layout of the circuit board for the power pack. C1 is large and is mounted above the other components (see Figure 12.5)

the negative rail (terminal pin J12) and the positive output terminal (F14). Switch on the main switch. The voltmeter should show a reading between 8.2V and 10.8V, depending on the setting of VR1. It may be a little outside this range if the values of the resistors R2, R3 and VR1 differ significantly from their marked values, as may occasionally happen. If the voltmeter shows a reading *much* outside this range, switch off and check for faulty wiring (did you remember to cut the copper strip at H10?). Alter the setting of VR1 and notice the corresponding changes in the voltmeter reading. Finally, adjust VR1 until the reading is exactly 9V.

Connect a small load across the output pins – a 1kΩ resistor will do. The output should remain unchanged, at 9V. Next try the effect of a larger load – one approaching the maximum that the circuit is designed to supply. A 27Ω resistor passes a current of 333mA; connect such a resistor across the output terminals. The output should still remain steady at exactly 9V. A resistor passing 333mA at 9V is dissipating power at the rate of 3W, so it will soon become too hot to hold! Unless it is a wire-wound resistor designed to work under such conditions it may soon become damaged by the heat; leave it connected only long enough for you to observe that the output remains unchanged by the low-resistance load. You should next try resistors of even lower value. A 22Ω resistor will pass 409mA, and may cause the output to drop by about 0.1 or 0.2V, which is not a serious problem. A 18Ω resistor would pass 500mA at 9V, which is more than the circuit is designed to deliver. Try putting an 18Ω resistor across the terminals; the output should fall immediately to about 8V. The current-limiting action is coming into play. At 8V an 18Ω resistor passes 444mA – just about the maximum possible current. With a 10Ω resistor the output falls to 4.4V – again, the unit is delivering its maximum current, 440mA. Your unit should show behaviour similar to that described above unless you have decided to limit the current to some other value by employing a resistor of some other value for R1.

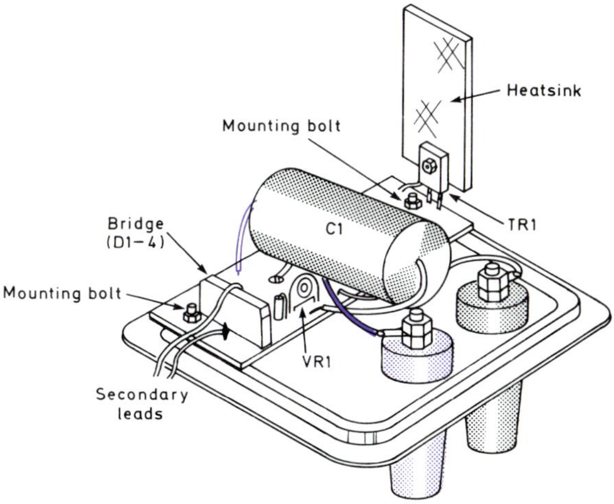

Figure 12.5

Assembly of the circuit board and terminals below the lid of the case, inverted in this drawing

If the circuit is working correctly, it may now be mounted on the lid of the tin as in Fig. 12.5. It is mounted on two 6B.A. bolts. The two bolts are passed through holes drilled in the lid and after placing a shake-proof washer on each bolt a nut is tightened on each. Then a second nut is threaded onto each to act as a spacer, holding the board clear of the lid. The board is then placed over the bolts, and a second washer and third nut used to hold the board in position. Fig. 12.5 shows a heat sink bolted to TR1. This is cut from a piece of 18 swg aluminium. Make it as large as will fit comfortably into the space available. It can, if necessary, be bent to avoid contact with exposed wiring or the metal parts of the transformer. Check this very carefully, as you press the lid into position, for the metal face is electrically connected to the collector of TR1 and is at a potential of + 18V.

Fig. 12.5 shows two large screw terminals being used as output terminals for the unit. The type shown can accept 4mm plugs, spade terminals, or bare ends of wires. These terminals are particularly useful if you intend to use the unit for providing power for electronic projects you are building and testing. They can also provide a connection to electrical equipment through leads terminated with press studs or crocodile clips. In the prototype an additional output socket was provided to allow for the powering of equipment that normally operates from PP3 batteries. This can be seen just behind the main terminals in

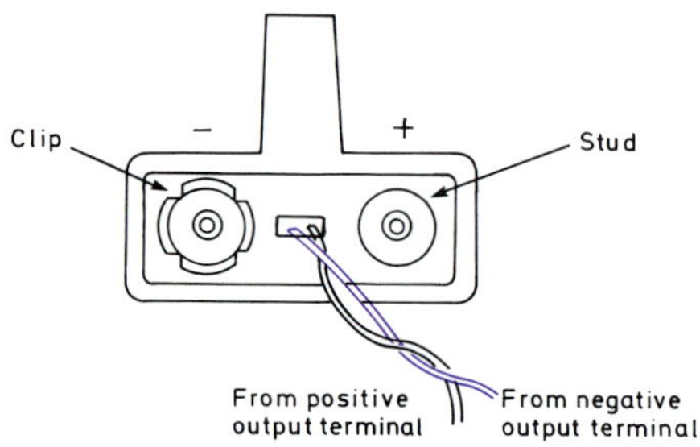

Figure 12.6

Connections to a battery clip wired as a *source* of current from the power pack

Fig. 12.2. The socket is a 2-pin socket and the plug that fits into it is non-reversible. This is wired to a battery connector with twin studs, suited for connection to a PP3 battery. The connector is wired so that the stud (the one that is gripped) is connected to the positive terminal of the unit, and the clip that grips the stud is wired to the negative terminals, as in Fig. 12.6. In this way, it has the same polarity as the PP3 battery it is replacing. Note that if you buy a ready wired battery connector, the red wire is connected to the clip, for this is the one that normally *clips on to* the positive stud battery. This red wire must be connected to the *negative* terminal of the power pack, for here we are using the battery connector as a *source* of power, not as a *receiver* of power from a battery.

Before connecting this unit to any electrical appliance it is essential to check that the appliance is designed to operate from a 9V supply.

Appliances that use a PP3 or PP6, or that use six 1.5V cells in series (HP2, HP11, etc) may normally be operated from this unit.

Variations on the circuit

Voltages up to 15V

The equation of p. 78 shows that if $(R_a + R_b) = 10.5\text{k}\Omega$, then R_b must be 4.9kΩ. To achieve this, we use resistors of different values: R2 = 4.7kΩ, VR1 = 2.2kΩ and R3 = 3.3kΩ. This gives a total 10.2kΩ, and R_b must therefore be 4.76kΩ. Since R3 is 3.3kΩ, the resistance between the wiper of VR1 and its negative end must be 4.76 – 3.3 = 1.46kΩ, and this falls within the 2.2kΩ range of this potentiometer.

Voltages above 15V

A transformer giving higher output voltage is required, and C1 must be rated to operate at a correspondingly higher working voltage. The 723 has a maximum input voltage of 40V and a maximum output voltage of 37V. The required values of R2, VR1 and R3 can be calculated by using the equation on p. 78.

Lower voltages

To obtain output voltages *lower* than the reference voltage (7V), a different arrangement of resistors is required (Fig. 12.7). Now it is the

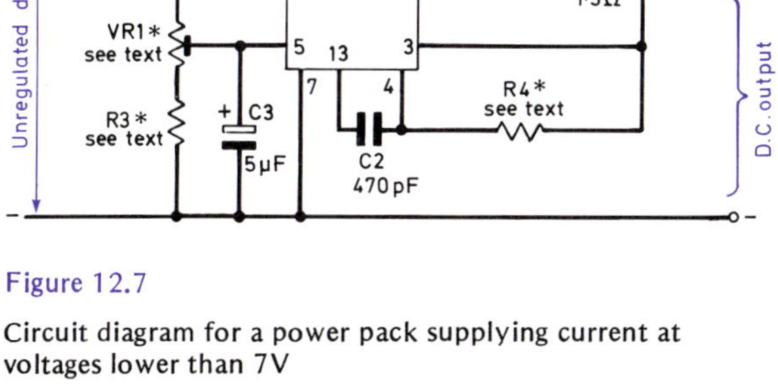

Figure 12.7

Circuit diagram for a power pack supplying current at voltages lower than 7V

reference voltage that is being applied to a potential divider. Instead of dividing the output voltage to match the reference voltage, we are now dividing the reference voltage to match the output. For example, to obtain an output of 6V, we must arrange for the *wiper* potential to be 6V.

To calculate R_b we need a slightly different equation:

$$\frac{V_o}{V_{ref}} = \frac{R_b}{R_a + R_b}$$ (contrast this with the equation on p. 78)

If $R_a + R_b = 10k\Omega$, we can calculate R_b

$$R_b = \frac{V_o}{V_{ref}} \times (R_a + R_b) = \frac{6}{7} \times 10\,000 = 8\,571\Omega$$

To allow this setting, we could make R2 = 1kΩ, VR1 = 2.2kΩ and R3 = 6.8kΩ. These values would also allow an output of 5V to be obtained, so making the unit ideal for driving TTL logic circuits.

The value of R4 (Fig. 12.7) is not critical, but to minimise temperature drift it should have a value close to that calculated by this equation:

$$R4 = \frac{R_a \times R_b}{R_a + R_b}$$

In the example quoted for a 6V supply,

$$R4 = \frac{1429 \times 8571}{10\,000} = 1\,225\Omega$$

A resistor of 1.2kΩ should therefore be used for R4. By using a configuration of the kind shown in Fig. 12.7 it is possible to obtain output voltages in the range 2V to 7V. The modification to the layout of the circuit board is shown in Fig. 12.8.

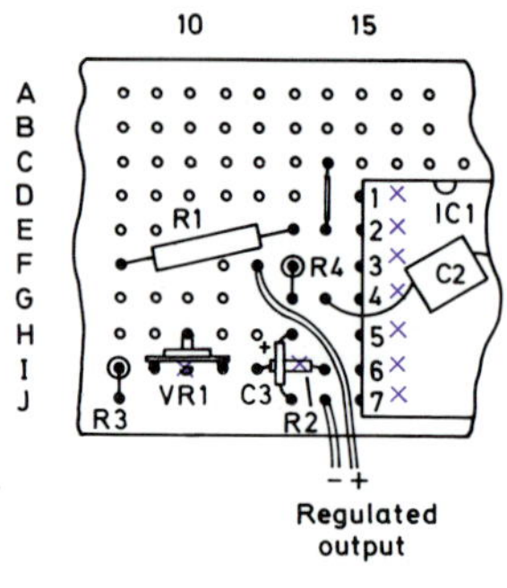

Figure 12.8

Layout of the circuit board for the power pack shown in Figure 12.7

Variable voltages

Any of the circuits described in this chapter may be used as the basis of a variable-voltage power pack. You could build a low-voltage power pack, to cover the range 2V to 7V, or a high-voltage power pack to cover a range from 7V to 15V or even higher. You would probably wish to build such a unit into a larger case, to allow yourself a little more room to manipulate the soldering iron, and perhaps to allow a voltmeter to be mounted on the front panel to indicate the output voltage being delivered. Instead of using a preset potentiometer for VR1, terminal pins are inserted in the board at holes G10, H9 and H11, and leads run from these to a potentiometer mounted on the front panel.

Small currents

If you will not need to take more than 150mA from the power pack, this can be obtained *directly* from pin 10 of the 723. There is no need

Table 13.1. Components list for basic power pack

Resistors	
R1	1.5Ω wirewound, 2.5W
R2	1.5kΩ carbon, ¼W
R3	6.8kΩ carbon, ¼W
Potentiometer	
VR1	2.2kΩ miniature vertical potentiometer
Capacitors	
C1	4 700μF, electrolytic 25V
C2	470pF ceramic
Semiconductors	
TR1	BD131
D1–4	Bridge rectifier type B40C1500 (A. Marshall Ltd)
Integrated circuit	
IC1	723 variable voltage regulator
Miscellaneous	
T1, transformer, mains voltage primary, 9-0-9 secondary, rated 0.5A or more; LP1 panel-mounting neon indicator lamp with integral resistor; DPST toggle switch; 0.1in matrix stripboard 10 strips by 24 holes; terminal pins for 0.1in matrix board (four off) 4 B.A. bolts, nuts and shakeproof washers (two off); 4 B.A. solder tag; 6 B.A. bolts (three off); 6 B.A. nuts (seven off); 6 B.A. shakeproof washers (five off); terminal sockets, 4mm (1 red, 1 black) or other output sockets; PP3 battery connector (optional); scrap 18swg aluminium sheet; connecting wire; mains lead and plug; solder	

for the power transistor. R1 should have the value 4.7Ω to limit current to 150mA, but need not be a wire-wound resistor, since power dissipation is low. The circuit board layout is the same as shown in Fig. 12.4 or 12.8, except that TR1 is omitted, and the connecting wire between E14 and C14 should now run from E4 to H19, so connecting pin 10 directly to pin 2.

Appendix

Handling CMOS integrated circuits

CMOS i.c.s can be damaged by electrostatically-produced charges residing on clothes, fingers, the soldering iron and other tools. Damage is normally avoided by observing these precautions:

(1) Keep the i.c. in its original container or wrapping until required for use. CMOS i.c.s are normally supplied with their pins interconnected by embedding them in a scrap of *conductive* plastics foam (usually black). Alternatively, they may be supplied wrapped in metal foil or inserted in short lengths of specially shaped aluminium channelling. All these methods short circuit the pins of the i.c. together.

(2) Work on a metal surface, such as a sheet of aluminium kitchen foil or an unpainted biscuit-tin lid. The surface must be earthed by wiring it to mains earth, or to a cold-water pipe, or to a strip of metal embedded in damp soil outdoors.

(3) Do not wear clothes made of synthetic fibre (nylon, terylene etc.) while working with i.c.s. Roll up your sleeves and, whenever possible, rest your wrist or fore-arm on the metal surface on which you are working. Touch your finger-tips to the sheet before each operation on the i.c.

(4) Earth the soldering iron and any other tools or instruments that may make contact with the IC or circuit board during construction. Touch the tip of the soldering iron briefly against the metal surface before each operation.

(5) The use of an IC socket makes it necessary to observe the above precautions only during the actual insertion of the i.c. into the socket. Once the i.c. has been inserted in its socket or fully soldered into the board and *providing all other components have already been correctly assembled,* the circuit board may be handled normally.

(6) Do not insert i.c.s into sockets or remove them from sockets, while the power supply is on.

(7) Never apply any voltage input to the input terminals pins of an i.c. unless the power supply is connected to pins 7 and 14 and is already on. This is important when testing circuits containing CMOS i.c.s. Conversely, disconnect input voltages before switching off the power supply to the i.c.

(8) Finally, observe the precaution that applies *to all i.c.s being soldered directly* to the board – solder the pins one at a time, and allow a minute or so between pins, so that the i.c. does not become overheated.

Dimensions of stripboard

The sizes of boards have been specified in the components lists by quoting the number of strips and holes. These are the most important facts from the point of view of the constructor. However, most suppliers specify board size by their dimensions in inches. To assist you in ordering boards of the correct size, the table below relates dimensions in inches to the number of holes and strips.

Length of side (inches)	*Number of strips or holes*	
	0.15in matrix	*0.1in matrix*
2.5	16	24
3.75	25	36
5	33	50

If you are intending to purchase a large board to cut into smaller pieces (this is an economical thing to do), remember that it is usual to cut the board along a row of holes. You *lose* one row of holes (or a strip) for each cut you make. For example, a board 3.75in × 5in (0.1in matrix) yields one board 18 strips by 50 holes and one board *17* strips by 50 holes. Boards are often a tight fit in the slots provided in plastics moulded cases. Consequently, you may need to file away the ends of a board to make it a reasonably loose fit. This may entail loss of a row of holes at the end of the board.